T0245145

LONDON MATHEMATICAL SOCIETY STUDENT TEXTS

Managing editor: Professor C.M. Series, Mathematics Institute
University of Warwick, Coventry CV4 7AL, United Kingdom

London Mathematical Society Student Texts 31

The Laplacian on a Riemannian Manifold

An Introduction to Analysis on Manifolds

Steven Rosenberg
Boston University

CAMBRIDGE
UNIVERSITY PRESS

PUBLISHED BY THE PRESS SYNDICATE OF THE UNIVERSITY OF CAMBRIDGE
The Pitt Building, Trumpington Street, Cambridge CB2 1RP, United Kingdom

CAMBRIDGE UNIVERSITY PRESS
The Edinburgh Building, Cambridge, CB2 2RU, United Kingdom
40 West 20th Street, New York, NY 10011-4211, USA
10 Stamford Road, Oakleigh, Melbourne 3166, Australia

© Cambridge University Press 1997

First published 1997
Reprinted 1998

A catalogue record for this book is available from the British Library

ISBN   0 521 46300 9 hardback
ISBN   0 521 46831 0 paperback

Transferred to digital printing 2003

Contents

Introduction

From the basic definitions, differential topology studies the global properties of smooth manifolds, while differential geometry studies both local properties (curvature) and global properties (geodesics). This text studies how differential operators on a smooth manifold reveal deep relationships between the geometry and the topology of the manifold. This is a broad and active area of research, and has been treated in advanced research monographs such as [5], [30], [59]. This book in contrast is aimed at students knowing just the basics of smooth manifold theory, say through Stokes' theorem for differential forms. In particular, no knowledge of differential geometry is assumed.

The goal of the text is an introduction to central topics in analysis on manifolds through the study of Laplacian-type operators on manifolds. The main subjects covered are Hodge theory, heat operators for Laplacians on forms, and the Chern-Gauss-Bonnet theorem in detail. Atiyah-Singer index theory and zeta functions for Laplacians are also covered, although in less detail. The main technique used is the heat flow associated to a Laplacian. The text can be taught in a one year course, and by the conclusion the student should have an appreciation of current research interests in the field.

We now give a brief, quasi-historical overview of these topics, followed by an outline of the book's organization.

The only natural differential operator on a manifold is the exterior derivative d taking k-forms to $(k + 1)$-forms. This operator is defined purely in terms of the smooth structure. Using d, we can define de Rham cohomology groups, the Euler characteristic and the degree of a map of smooth manifolds, all of which give topological information [32]. With some more work, we can reformulate intersection theory in terms of integration of closed forms [11], and so in principle determine the entire real cohomology ring of the manifold.

Once we enter the domain of differential geometry by introducing a Riemannian metric on the manifold, we find a series of differential operators Δ^k, the Laplacians on k-forms, associated to the metric. In particular, the Laplacian on functions generalizes the usual Laplacians on $\mathbf{R}^n$ and on the circle.

On a compact manifold, the spectrum $\{\lambda_i^k\}$ of Δ^k contains both topological and geometric information. In particular, by the Hodge theorem the dimension of the kernel of Δ^k equals the kth Betti number, and so the Laplacians determine the Euler characteristic χ.

The geometric information contained in Δ^k is more difficult to extract. We consider the heat equation $(\partial_t + \Delta^k)\omega = 0$ on k-forms with solution given by the heat semigroup $e^{-t\Delta^k}\omega_0$, ω_0 being the initial k-form. The behavior of the trace of the heat semigroup, $\text{Tr}(e^{-t\Delta^k}) = \sum_i e^{-\lambda_i^k t}$, as $t \to 0$ is controlled by an infinite sequence of geometric data, starting with the volume of the manifold and the integral of the scalar curvature. This is surprising, since the trace is constructed just from the spectrum of Δ^k.

Now the kernel of Δ^k controls the behavior of $e^{-t\Delta^k}$ as $t \to \infty$. It turns out that the sum $\sum_k (-1)^k \text{Tr}(e^{-t\Delta^k})$ of the traces of the heat kernels is independent of t. The long time behavior of this sum equals the Euler characteristic, while the short time behavior is given by an integral of a complicated curvature expression.

If the dimension of the manifold M is two, this equality of long and short time behavior of the heat flow leads to the Gauss-Bonnet theorem: $\chi(M) = \int_M K\, dA$, where K is the Gaussian curvature and dA is the area element. Note the remarkable fact that the integrand is independent of the Riemannian metric. Of course, there are much simpler proofs of Gauss-Bonnet, but this technique shows that a generalization of Gauss-Bonnet exists in higher dimensions. The explicit determination of the curvature integrand in this generalization, originally due to Chern by other methods, is one of the main results of the text. The proof is a modification of techniques introduced by Getzler around 1985.

The Chern-Gauss-Bonnet theorem, first shown around 1945, can itself be generalized. The Riemannian metric induces Hilbert space structures on the spaces of k-forms, and so d has an adjoint δ taking $(k + 1)$-forms to k-forms. Recall that the index of an operator D on a Hilbert space is given by $\text{ind}(D) = \dim \ker (D) - \dim \text{coker} (D)$ when the kernel and cokernel are finite dimensional. From Hodge theory, we find that the index of the first order geometric operator $d + \delta$ taking even forms to odd forms is just the Euler characteristic.

This suggests that we look for other geometrically defined operators whose index is a topological invariant. One example is the signature operator, whose index, the signature, is an important topological quantity associated to the middle dimensional cohomology of the manifold. The heat equation approach again gives the signature as the integral of a curvature expression. Even more generally, the Atiyah-Singer index theorem, dating from the early 1960s, shows that the index of any elliptic first order geometric operator D is given by such an integral, even though the index need not have an obvious topological interpretation. Thus we can state the Atiyah-Singer index theorem schematically as $\text{ind}(D_g) = \int_M \mathcal{R}(g)$, where g denotes a Riemannian metric and $\mathcal{R}(g)$ denotes the curvature expression.

The index of these operators will be independent of the Riemannian metric, and so $\int_M \mathcal{R}(g)$ is also metric independent, as in the Gauss-Bonnet theorem. This implies that these particular curvature integrands represent the same cohomology class. As a result, Chern-Weil theory, which constructs representatives of certain cohomology classes from a Riemannian metric, naturally enters the picture.

To summarize, the Gauss-Bonnet theorem equates the topological quantity

$\chi(M)$ with the geometric quantity $\int_M K \, dA$. At the end of the generalization process, the Atiyah-Singer index theorem equates the analytic quantity ind(D_g) with a topological quantity given by Chern-Weil theory. In fact, the index theorem applies to all elliptic operators, not just geometrically defined operators. This reinterpretation of the nature of both sides of the index theorem indicates the depth of this theory.

To see what lies beyond index theory, we go back to differential topology and first ask what lies beyond cohomology. If the (twisted) cohomology groups of a manifold vanish, a subtler or secondary topological invariant, the Reidemeister torsion, is well defined. In the 1970s, concurrent with the development of the heat equation approach to index theory, Ray and Singer proposed an analytic analogue of Reidemeister torsion defined from the Laplacians on k-forms. The definition involves the zeta function of the Laplacian, which encodes the spectrum of the Laplacian differently from the trace of the heat operator. This analytic torsion was shown to equal the Reidemeister torsion around 1980. Recent work of Bismut and Lott [8] has clarified the connection between index theory and analytic torsion.

Index theory and analytic torsion continue to develop in many directions, including K-theory, operator theory, number theory and mathematical physics. (For example, in operator theory it is now considered hopelessly old fashioned to think of the index as an integer.) Hopefully, readers of this book will contribute to this development.

In more detail, Chapter 1 treats the heat equation approach to Hodge theory. We discuss heat flow for the Laplacian on $\mathbf{R}$ and the circle. Riemannian metrics are defined, as are the associated Hilbert spaces of k-forms. The Laplacians on forms are given in terms of the Riemannian metric. After proving the basic analytic results (Sobolev embedding theorem, Rellich compactness theorem), we give a heat equation proof of the Hodge theorem, which gives an eigenform decomposition of the Hilbert spaces of forms generalizing Fourier series on the circle. This proof assumes the existence of an integral kernel, the heat kernel, for heat flow for the Laplacians on forms; the construction of the heat kernel is in Chapter 3. The proof shows that the long time behavior of the heat flow is controlled by the kernel of the Laplacian. We then use Gårding's inequality to prove the standard regularity results for the Laplacians and to give the Hodge decomposition of the spaces of smooth forms. (The more standard elliptic/potential theoretic proof of the Hodge theorem is given in the exercises.) We define the de Rham cohomology groups and show that the kernel of the Laplacian on k-forms is isomorphic to the kth de Rham cohomology group. Thus the long time heat flow is controlled by the topology of the manifold.

Chapter 2 covers just those parts of differential geometry needed to construct the heat kernel. We introduce the various curvatures associated to a Riemannian metric and prove that the Riemann curvature tensor is the obstruction to a metric being locally flat. We define the Levi-Civita connection for a Riemannian metric, and prove the Bochner formula relating the Laplacian on forms to this connection and the Riemannian curvature. The Bochner formula is proved using supersymmetry/fermion calculus methods, and it leads to a quick proof of

Gårding's inequality used in Chapter 1. We then turn to the study of geodesics and the exponential map, and give a technical computation of the Laplacian on functions in Riemannian polar coordinates.

In Chapter 3, we construct the heat kernel for the Laplacians on functions and forms. We use Duhamel's formula to motivate the complicated calculations. The construction shows that the short time behavior of the heat flow is determined by the geometry of the Riemannian metric. We also show that the heat kernel on functions is positive.

Chapter 4 discusses the heat equation approach to Atiyah-Singer index theory. The main idea is to compare the long time and short time information in the heat flows. We first show that the Euler characteristic is given by an integral of a curvature expression generalizing the Gauss-Bonnet theorem for surfaces. Following unpublished work of Parker [53], we give a fermion calculus proof of the Chern-Gauss-Bonnet theorem by showing that the integrand is the expected Pfaffian of the curvature. There is a brief discussion of Chern-Weil theory, showing how characteristic classes have representative forms constructed from the curvature of a Riemannian metric. This allows us to give a precise formulation of the Hirzebruch signature theorem. We do not prove this theorem, as the fermion calculus is more involved, but refer the reader to proofs in more advanced texts. We briefly discuss the Hirzebruch-Riemann-Roch theorem; this discussion assumes familiarity with complex geometry and can be omitted. (Since the Dirac operator is even trickier to define and is treated extensively in [5], [30], [59], we do not discuss it at all.) Finally, we define elliptic operators and state the Atiyah-Singer index theorem.

In Chapter 5, we discuss the zeta function of the Laplacian on forms. We show that the poles and special values of the zeta function contain the same information as the short time asymptotics of the heat kernel. For the conformal Laplacian on functions, we show that $\zeta(0)$ is a conformal invariant given by a curvature expression, and that $\zeta'(0)$ is a subtler conformal invariant. We then digress to give Sunada's elegant construction of nonhomeomorphic manifolds whose Laplacians on functions have the same spectrum. We define Reidemeister and analytic torsion, which involves interpreting the determinant of the Laplacian in terms of $\zeta'(0)$. We show that analytic torsion is independent of the Riemannian metric, and so gives a smooth invariant of the manifold. We finish with a (not self-contained) discussion of the recent work of Bismut and Lott, which states precisely in what sense analytic torsion arises as a secondary invariant when no information is available from index theory techniques.

I would like to acknowledge the hospitality of Keio University and the University of Warwick, where much of this text was written. Special thanks are due to Tom Parker for explaining his proof of the Chern-Gauss-Bonnet theorem, to Eric Boeckx for a careful reading of the text, and to the NSF and JSPS for their support. A much different form of support was provided by my wife, Sybil, and an extremely different form by my children, Sam and Selene. This book is dedicated to my family.

Chapter 1

The Laplacian on a Riemannian Manifold

In this chapter we will generalize the Laplacian on Euclidean space to an operator on differential forms on a Riemannian manifold. By a Riemannian manifold, we roughly mean a manifold equipped with a method for measuring lengths of tangent vectors, and hence of curves. Throughout this text, we will concentrate on studying the heat flow associated to these Laplacians. The main result of this chapter, the Hodge theorem, states that the long time behavior of the heat flow is controlled by the topology of the manifold.

In §1.1, the basic examples of heat flow on the one dimensional manifolds S^1 and $\mathbf{R}$ are studied. The heat flow on the circle already contains the basic features of heat flow on a compact manifold, although the circle is too simple topologically and geometrically to really reveal the information contained in the heat flow. In contrast, heat flow on $\mathbf{R}$ is more difficult to study, which indicates why we will restrict attention to compact manifolds. In §1.2, we introduce the notion of a Riemannian metric on a manifold, define the spaces of L^2 functions and forms on a manifold with a Riemannian metric, and introduce the Laplacian associated to the metric. The Hodge theorem is proved in §1.3 by heat equation methods. The kernel of the Laplacian on forms is isomorphic to the de Rham cohomology groups, and hence is a topological invariant. The de Rham cohomology groups are discussed in §1.4, and the isomorphism between the kernel of the Laplacian and de Rham cohomology is shown in §1.5.

Before we start, we note that while the simplest differential operator d/dt on the real line generalizes to the exterior derivative d on a smooth manifold, it is not possible to generalize the second derivative to manifolds without the additional structure of a Riemannian metric. Thus it can be argued that the Laplacian is the simplest, and hence the most basic, differential operator on functions on a Riemannian manifold. Just as the study of the exterior derivative leads to important results, such as de Rham's theorem, relating the smooth structure of the manifold to its underlying topological structure, the study of the Laplacian leads to even deeper results, such as the geometric version of the Atiyah-Singer index theorem, which relates the topology, the smooth structure, and the geometry of a Riemannian manifold.

Notation: Given a smooth map $f\colon M \to N$ between manifolds, we will denote the differential of f by either df or f_*.

1.1 Basic Examples

Because the theory of the Laplacian on a Riemannian manifold involves some technical preliminaries, we begin by examining some simple examples. In fact, considering the Laplacian and the associated heat flow on just S^1 and $\mathbf{R}$ highlights essential differences between the Laplacian on a compact and on a noncompact manifold.

First, recall that if $T : V \to V$ is a symmetric, nonnegative linear transformation of a finite dimensional inner product space V, then there exists an orthonormal basis of eigenvectors of V with eigenvalues $0 \leq \lambda_1 \leq \ldots \leq \lambda_n$. The set $\{\lambda_i\}$ is called the *spectrum* of T, denoted $\sigma(T)$, and is characterized by the property

$$\lambda \notin \sigma(T) \Leftrightarrow (T - \lambda I)^{-1} \text{ exists} \Leftrightarrow \mathrm{Ker}(T - \lambda I) = 0.$$

This eigenvector decomposition of V generalizes to the infinite dimensional case where V is a Hilbert space and T is a compact operator, i.e. an operator such that if $\{v_i\}$ is a bounded sequence in V, then $\{Tv_i\}$ has a convergent subsequence. (For example, any projection onto a finite dimensional subspace is compact, and in fact any compact operator is the norm limit of such finite rank operators.) In this case, the spectral theorem for compact operators says that V again has an orthonormal basis of eigenvectors for T, each eigenspace has only finite multiplicity, and the only (finite or infinite) accumulation point for the set of eigenvalues is zero. In particular, since the absolute values of the eigenvalues are bounded, the operator T is itself bounded. Remember that in infinite dimensions a linear operator may well be unbounded, or equivalently discontinuous.

The spectral theorem for compact operators is an easy generalization of the finite dimensional situation. We want to show that this eigenvector decomposition holds for certain *unbounded* differential operators on compact manifolds. The space V will be some Hilbert space of functions or forms on the manifold. We remark that unbounded operators are only defined on a dense subset of a Hilbert space, and in general one must be very careful to define the domains of such operators and their adjoints correctly. The domains of definition of our unbounded operators are rather easy to construct on compact manifolds, but noncompact manifolds are more difficult to treat. We will follow the standard practice of glossing over these problems, but here are some references for the reader: for unbounded operators in general [70] is quite thorough, while the domains of various Laplacians are stated carefully in [26].

1.1.1 The Laplacian on S^1 and R

The first example to consider is the circle. We set $V = L^2(S^1) = L^2(S^1, \mathbf{C})$, the space of complex valued L^2 functions. (We could just as easily deal with real valued functions, but the notation is a bit more involved.) The simplest differential operator on the circle is of course $d/d\theta$. However, this operator generalizes on a manifold to the exterior derivative $d : \Lambda^0 M \to \Lambda^1 M$, which takes one space to a different space and hence does not have a spectrum; only on one dimensional manifolds can one identify one-forms with functions, e.g. by identifying $f(\theta)d\theta$ with $f(\theta)$. The next simplest operator is the Laplacian or second derivative,

$$\Delta = -\frac{d^2}{d\theta^2},$$

where once and for all we adopt the geometers' convention of placing a minus sign in the definition. We will see that this operator does generalize naturally to an operator on a manifold taking functions again to functions.

Exercise 1: *We might try to consider $d : \Lambda^* \to \Lambda^*$ as an operator taking forms of mixed degree to forms of mixed degree, in which case $\sigma(d)$ is well defined. Show that for any manifold the only eigenvalue is zero, that zero has infinite multiplicity, and that there is an infinite dimensional space of forms which are not in the zero eigenspace. Although we have not yet made Λ^* into a Hilbert space, this shows that a nice spectral decomposition of Λ^* with respect to d does not exist.*

The eigenfunction decomposition of $L^2(S^1)$ is quite well known. An orthonormal basis is given by the trigonometric polynomials $\{e^{in\theta}\}$, $n \in \mathbf{Z}$, and $\Delta e^{in\theta} = n^2 e^{in\theta}$. Thus $L^2(S^1)$ decomposes into eigenspaces with eigenvalues $\{n^2 : n = 0, 1, ...\}$ and each eigenspace has multiplicity two, except for the eigenspace of zero, which has multiplicity one. Notice that the eigenfunction decomposition of $f \in L^2$ is given by

$$f = \sum_n a_n e^{in\theta} = \sum_n \langle f, e^{in\theta} \rangle e^{in\theta},$$

with

$$\langle f, g \rangle = \frac{1}{2\pi} \int_{S^1} f(\theta)\overline{g(\theta)} \, d\theta$$

the usual L^2 inner product, which is just the Fourier series decomposition of f. Note also that

$$||e^{in\theta}/n|| \to 0, \text{ but } ||\Delta(e^{in\theta}/n)|| \to \infty$$

as $n \to \infty$, so Δ is unbounded.

The (formal) theory of Fourier series is quite old, dating from around 1825. By the end of the 19th century, Sturm-Liouville theory provided a powerful generalization of Fourier series. This theory typically treats operators of the form

$$D = -\frac{d^2}{dt^2} + A(t)\frac{d}{dt} + B(t),$$

for certain smooth functions A and B, acting on $L^2[-\pi, \pi]$ with certain boundary conditions. In particular, if periodic boundary conditions are imposed, we are working on $L^2(S^1)$. According to this theory, all such operators D give an eigenfunction decomposition of $L^2[-\pi, \pi]$. Moreover, the eigenspaces are finite dimensional, and the eigenvalues $\{\lambda_n\}$ accumulate only at ∞. However, for general A and B it will not be possible to determine the corresponding eigenfunctions and eigenvalues. This situation is typical for Laplacian-type operators on compact manifolds.

We now define and compute the spectrum of $\Delta = -d^2/dt^2$ on $L^2(\mathbf{R})$. (This discussion is more advanced than the circle case, and can be skipped as it is not needed later.) For motivation, choose $\lambda \notin \sigma(D)$, where D is an operator of Sturm-Liouville type. Then not only does $(D - \lambda I)^{-1}$ exist, it is also a bounded operator, since

$$\sigma((D - \lambda I)^{-1}) = \{(\lambda_n - \lambda)^{-1}\}$$

is a bounded set. Moreover, as the reader should check for Δ, by the eigenfunction decomposition it can be shown that the range of $D - \lambda I$ is dense in $L^2(S^1)$, and so $(D - \lambda I)^{-1}$ extends to a bounded operator on all of L^2.

This (I hope) justifies the following definition of the spectrum of an unbounded operator; a complete justification is given by the spectral theorem for unbounded operators.

Definition: *Let D be a symmetric unbounded operator on a Hilbert space H. The spectrum of D, $\sigma(D) \subset \mathbf{R}$, is defined by the condition $\lambda \notin \sigma(D)$ iff $(D - \lambda I)^{-1}$ can be extended to a bounded operator on all of H. Equivalently, $\lambda \notin \sigma(D)$ iff (i) $\mathrm{Ker}(D - \lambda I) = 0$, (ii) the image of $D - \lambda I$ is dense, and (iii) on the image, $(D - \lambda I)^{-1}$ is bounded.*

Now set $D = \Delta$ acting on $H = L^2(\mathbf{R})$. We first consider $\lambda \geq 0$. If we look for eigenfunctions, we find that the only solutions to $\Delta f = \lambda f$ are $f(x) = \exp(\pm i\sqrt{\lambda}x)$, for any $\lambda \in \mathbf{R}^+ \cup 0$. However, none of these eigenfunctions is in L^2. This does not mean that $\sigma(\Delta) = \emptyset$. Consider a function $\psi_N(x)$ on $\mathbf{R}$ which satisfies $\psi_N \geq 0$ and

$$\psi_N(x) = \begin{cases} 0 & \text{if } x \in (-\infty, -N] \cup [N, \infty), \\ 1 & \text{if } x \in [-N+1, N-1]. \end{cases}$$

For fixed $\lambda \geq 0$, there exist constants $C, C' > 0$ such that

$$\|(\Delta - \lambda I)(\psi_N e^{i\sqrt{\lambda}x})\| \leq C \leq \frac{C'}{N}\|\psi_N e^{i\sqrt{\lambda}x}\|. \tag{1.1}$$

For ψ_N is nonconstant only on $[-N, -N+1] \cup [N-1, N]$, so the function on the left hand side of (1.1) is zero except on this interval.

Exercise 2: *(i) Show that (1.1) implies $\|(\Delta - \lambda I)^{-1}\| = \infty$.*

(ii) Show that $\lambda < 0 \Rightarrow \lambda \notin \sigma(\Delta)$. Hint: Show that there are no L^2 eigen-functions for λ. If the image of $\Delta - \lambda I$ is not dense, take a non-zero $\alpha \in L^2(\mathbf{R})$ with $\alpha \perp \mathrm{Im}(\Delta - \lambda I)$. Show that

$$0 = \int_{\mathbf{R}} \alpha(x)(\Delta - \lambda I)f(x)$$

for all compactly supported functions f. Thus α is a distributional or weak solution to $(\Delta - \lambda I)\alpha = 0$. (That is, if α were in $L^2(\mathbf{R})$, then we could integrate by parts and let f be a bump function to conclude that $(\Delta - \lambda I)\alpha = 0$.) Classical elliptic regularity results for the Laplacian then imply that in fact $\alpha \in L^2(\mathbf{R})$ (cf. §1.3.4). This is a contradiction. Finally, show that $\|(\Delta - \lambda I)^{-1}\| \le |\lambda|^{-1}$. For this last step, let $g = (\Delta - \lambda I)^{-1}f$. Show that in the L^2 norm and inner product we have

$$\|f\|^2 = \langle \Delta g, \Delta g \rangle - 2\lambda\langle \Delta g, g \rangle + \lambda^2 \langle g, g \rangle \ge \lambda^2 \|g\|^2.$$

Here $\langle \Delta g, g \rangle \ge 0$ by an integration by parts. Thus $\|g\|^2/\|f\|^2 \le (\lambda^2)^{-1}$.

From this exercise, we see that in fact $\sigma(\Delta) = [0, \infty)$.

Since the spectrum does not consist of a discrete set, we cannot have a Fourier series decomposition as on S^1. However, note the analogy between the Fourier series decomposition on S^1,

$$f(\theta) = \sum_n \left(\frac{1}{2\pi} \int_{S^1} f(\psi)e^{-in\psi}\, d\psi \right) e^{in\theta}, \tag{1.2}$$

and the Fourier inversion formula on $\mathbf{R}$,

$$f(x) = \frac{1}{\sqrt{2\pi}} \int_{\mathbf{R}} \left(\frac{1}{\sqrt{2\pi}} \int_{\mathbf{R}} f(y)e^{-i\xi y}\, dy \right) e^{i\xi x}\, d\xi. \tag{1.3}$$

While we will prove the existence of a formula corresponding to (1.2) on any compact manifold, no such formula corresponding to (1.3) is known for general noncompact manifolds. (In fact, the existence of (1.3) reflects the fact that $\mathbf{R}$ is a semisimple Lie group.)

1.1.2 Heat Flow on S^1 and $\mathbf{R}$

Given an initial distribution $f(\theta) = f(0, \theta)$ of heat on S^1, considered to be perfectly insulated, the distribution $f(t, \theta)$ of heat at time t is (allegedly) governed by the heat equation

$$(\partial_t + \Delta)f(t, \theta) = 0. \tag{1.4}$$

This is quite easy to solve explicitly. If $f(t, \theta) = \sum_n a_n(t)e^{in\theta}$ is the Fourier decomposition for $f(t, \theta)$, with $a_n(0) = a_n$ the nth Fourier coefficient for f, then plugging the Fourier decomposition into (1.4) gives

$$0 = \sum_n (\dot{a}_n(t) + n^2 a_n(t))e^{in\theta}.$$

It follows that $a_n(t) = a_n e^{-n^2 t}$, and so

$$f(t,\theta) = \sum_n e^{-n^2 t} a_n e^{in\theta}.$$

Note that as $t \to \infty$, $f(t,\theta) \to a_0$, which is the average value of f. This fits with our intuition that as $t \to \infty$ the heat should reach a constant equilibrium state; since the circle is insulated, the equilibrium value should be the average of the initial heat distribution.

The situation on **R** is as expected more complicated. Given the initial heat distribution $f(0,x) = f(x) \in L^2$, set

$$f(t,x) = \frac{1}{\sqrt{4\pi t}} \int_{\mathbf{R}} e^{-\frac{(x-y)^2}{4t}} f(y)\, dy. \tag{1.5}$$

Exercise 3: *(i) Verify that for continuous functions f,*

$$(\partial_t + \Delta)f(t,x) = 0 \text{ and } \lim_{t \to 0} f(t,x) = f(x).$$

(ii) Fill in the details in the following derivation for $f(t,x)$. We will assume familiarity with the Fourier transform; see Lemma 1.17. Let $\hat{f}(t,\xi)$ denote the Fourier transform of f in the x variable only. Show that $(\partial_t + \Delta)f(t,x) = 0$ implies

$$-|\xi|^2 \hat{f}(t,\xi) = \partial_t \hat{f}(t,\xi).$$

Conclude that $\hat{f}(t,\xi) = \hat{f}(\xi)e^{-t|\xi|^2}$. Thus

$$\begin{aligned}
\hat{f}(t,\xi) &= \hat{f}(\xi)e^{-t|\xi|^2} = \hat{f}(\xi)e^{-|\xi\sqrt{2t}|^2/2}\\
&= \hat{f}(\xi)\{\frac{1}{\sqrt{2t}}e^{-|x|^2/4t}\}^{\check{}}(\xi)\\
&= \{f * (\frac{1}{\sqrt{2t}}e^{-|x|^2/4t})\}^{\check{}}(\xi),
\end{aligned}$$

where {...}˘ denotes the Fourier transform of the function in the braces, and the star in the last line denotes convolution. Taking inverse Fourier transform yields

$$f(t,x) = \frac{1}{\sqrt{4\pi t}} \int_{\mathbf{R}} e^{-(x-y)^2/4t} f(y)\, dy.$$

(iii) Similarly show that the integral kernel for the heat equation on $\mathbf{R}^n$ is

$$e(t,x,y) = \frac{1}{(4\pi t)^{n/2}} e^{-|x-y|^2/4t}.$$

Thus (1.5) provides a solution to the heat equation in the form of an integral kernel

$$e(t,x,y) = \frac{1}{\sqrt{4\pi t}} e^{-\frac{(x-y)^2}{4t}}, \tag{1.6}$$

which is a smooth function on $(0, \infty) \times \mathbf{R} \times \mathbf{R}$. Notice that for all $f \in L^2$, $f(t, x) \to 0$ as $t \to \infty$, in accordance with our intuition that the heat should "dissipate to $\pm\infty$." While the solution (1.5) looks different from that of S^1, note that on S^1

$$
\begin{aligned}
f(t, \theta) &= \sum_n e^{-n^2 t} \langle f, e^{in\theta} \rangle e^{in\theta} \\
&= \frac{1}{2\pi} \int_{S^1} \sum_n e^{-n^2 t} e^{in\theta} \overline{e^{in\psi}} f(\psi) \, d\psi.
\end{aligned}
$$

Thus heat flow on S^1 is also given by the integral kernel

$$
\sum_n e^{-n^2 t} e^{in\theta} \overline{e^{in\psi}},
$$

which is easily seen to be smooth on $(0, \infty) \times S^1 \times S^1$.

Exercise 4: *Prove the Weierstrass approximation theorem: given $f \in C_c(\mathbf{R}^n)$ (where C_c denotes compactly supported continuous functions) and $\epsilon > 0$, there exists a polynomial $p(x)$ such that $\|f - p\|_\infty < \epsilon$ on the support of f. Hint: Apply the heat flow to f, which is valid since $f \in L^2$. Then apply Taylor's theorem to the resulting smooth function.*

These two heat kernels are in fact closely related. First, we have to readjust the inner product on $L^2(S^1)$ by setting $\langle f, g \rangle = \int_{S^1} f(\theta) \overline{g(\theta)} \, d\theta$. (This is just the integration theory on S^1 induced by the local isometry $\mathbf{R} \to S^1$ given by $x \mapsto x \pmod{2\pi}$.) This changes the orthonormal basis to $\{e^{in\theta}/\sqrt{2\pi}\}$, but leaves the rest of the previous discussion unchanged. Let $e_{\mathbf{R}}, e_{S^1}$ denote the heat kernels on $\mathbf{R}, S^1$, respectively. Set

$$
\tilde{e}_{S^1}(t, \theta, \psi) = \sum_{n \in \mathbf{Z}} e_{\mathbf{R}}(t, \theta, \psi + 2\pi n), \tag{1.7}
$$

where on the right side of (1.7) we consider θ, ψ to run over any interval of length 2π in $\mathbf{R}$; this is well defined since $e_{\mathbf{R}}(t, x, y) = e_{\mathbf{R}}(t, x + k, y + k)$ for $k \in \mathbf{R}$. Intuitively, we expect that $\tilde{e}_{S^1} = e_{S^1}$, since heat can "get" from $\theta \in S^1$ to $\psi \in S^1$ by flowing around the circle any number of times in either direction, or equivalently by flowing from $\theta \in \mathbf{R}$ to any translate of ψ in $\mathbf{R}$. This is indeed the case:

Lemma 1.8 $\tilde{e}_{S^1} = e_{S^1}$. *Thus*

$$
e_{S^1}(t, \theta, \psi) = \sum_{n \in \mathbf{Z}} e_{\mathbf{R}}(t, \theta, \psi + 2\pi n).
$$

PROOF. It is immediate that $(\partial_t + \Delta)\tilde{e}_{S^1} = 0$, where Δ acts on θ. Also,

$$
\lim_{t \to 0} \int_{S^1} \tilde{e}_{S^1}(t, \theta, \psi) f(\psi) \, d\psi = \lim_{t \to 0} \int_{\mathbf{R}} e_{\mathbf{R}}(t, \theta, \psi) f(\psi) \, d\psi,
$$

where we extend f to be periodic on $\mathbf{R}$. It is easy to check that even though $f \notin L^2(\mathbf{R})$, we do have

$$\lim_{t \to 0} \int_{\mathbf{R}} e_{\mathbf{R}}(t, \theta, \psi) f(\psi) \, d\psi = f(\theta).$$

The following lemma, whose generalization to manifolds will be used later, finishes the proof of Lemma 1.8:

Lemma 1.9 *Let $a(t, \theta, \psi), b(t, \theta, \psi) \in C^\infty(\mathbf{R}^+ \times S^1 \times S^1)$ satisfy $(\partial_t + \Delta)a = (\partial_t + \Delta)b = 0$, and*

$$\lim_{t \to 0} \int_{S^1} a(t, \theta, \psi) f(\psi) \, d\psi = \lim_{t \to 0} \int_{S^1} b(t, \theta, \psi) f(\psi) \, d\psi = f(\theta),$$

for all $f \in L^2(S^1)$. Then $a(t, \theta, \psi) = b(t, \theta, \psi)$.

Here we make the convention that the Laplacian always acts on the first space variable of the kernel unless otherwise indicated. When necessary, we use the notation $\Delta_\theta, \Delta_\psi$, etc. to indicate on which variable the Laplacian acts.

PROOF. We first show that $a(t, \theta, \psi) = a(t, \psi, \theta)$; similarly, $b(t, \theta, \psi)$ is symmetric in the space variables. For fixed θ, ψ two integrations by parts in the variable μ yield

$$\begin{aligned}
0 &= \int_{S^1} \Delta_\mu a(t', \psi, \mu) \cdot a(t - t', \theta, \mu) - a(t', \psi, \mu) \cdot \Delta_\mu a(t - t', \theta, \mu) \\
&= \int_{S^1} -\partial_{t'} a(t', \psi, \mu) \cdot a(t - t', \theta, \mu) - a(t', \psi, \mu) \cdot \partial_{t'} a(t - t', \theta, \mu) \\
&= -\partial_{t'} \int_{S^1} a(t', \psi, \mu) \cdot a(t - t', \theta, \mu).
\end{aligned}$$

Thus if we abbreviate $\lim_{t \to 0} \int_{S^1} a(t, \theta, \mu) f(\mu)$ by $\int_{S^1} a(0, \theta, \mu) f(\mu)$, we have

$$\begin{aligned}
0 &= \int_0^t dt' \, \partial_{t'} \int_{S^1} a(t', \psi, \mu) \cdot a(t - t', \theta, \mu) \\
&= \int_{S^1} a(t, \psi, \mu) \cdot a(0, \theta, \mu) - \int_{S^1} a(0, \psi, \mu) \cdot a(t, \theta, \mu) \\
&= a(t, \psi, \theta) - a(t, \theta, \psi).
\end{aligned}$$

Now consider the integral

$$\int_0^t ds \, \partial_s \int_{S^1} a(s, \theta, \mu) b(t - s, \mu, \psi) \, d\mu. \tag{1.10}$$

This equals

$$\lim_{t' \to 0} \left[\int_{S^1} a(t - t', \theta, \mu) b(t', \mu, \psi) \, d\mu - \int_{S^1} a(t', \theta, \mu) b(t - t', \mu, \psi) \, d\mu \right],$$

which reduces to

$$a(t, \theta, \psi) - b(t, \theta, \psi).$$

On the other hand, (1.10) also equals

$$\int_0^t ds \left[\int_{S^1} \partial_s a(s, \theta, \mu) \cdot b(t - s, \mu, \psi) \, d\mu + \int_{S^1} a(s, \theta, \mu) \cdot \partial_s b(t - s, \mu, \psi) \, d\mu \right]$$

$$= \int_0^t ds \left[\int_{S^1} -\Delta_\theta a(s, \theta, \mu) \cdot b(t - s, \mu, \psi) \, d\mu + \int_{S^1} a(s, \theta, \mu) \cdot \Delta_\mu b(t - s, \mu, \psi) \, d\mu \right].$$

$$(1.11)$$

By the symmetry of $a(t, x, y)$ in x and y, we can replace $-\Delta_\theta a(s, \theta, \mu)|_{(s, \theta, \mu)}$ in (1.11) by $-\Delta_\mu a(s, \theta, \mu)|_{(s, \mu, \theta)}$. Two integrations by parts then replace the first integrand on the right hand side of (1.11) by

$$-a(s, \mu, \theta) \Delta_\mu b(t - s, \mu, \psi) = -a(s, \theta, \mu) \Delta_\mu b(t - s, \mu, \psi),$$

which cancels with the second integrand, finishing the proof.

We denote the operator taking the heat distribution f to the time t heat distribution $f(t, x)$ by

$$e^{-t\Delta} : L^2 \to L^2,$$

for heat flow on either S^1 or $\mathbf{R}$; the notation is suggested by the fact that $e^{-t\Delta}$ acts by multiplication by e^{-tn^2} on the n^2-eigenspace of Δ on S^1, and is justified by the spectral theorem for unbounded operators in the case of $\mathbf{R}$ (see [70]). The trace of the heat operator on S^1 is given by the "trace" of the heat kernel:

$$\operatorname{Tr} e^{-t\Delta} = \sum_n e^{-n^2 t} = \int_{S^1} e_{S^1}(t, \theta, \theta) \, d\theta.$$

For short time, the trace of the heat kernel on S^1 looks like the trace of the heat kernel on $\mathbf{R}$, in the sense that

$$\sum_n e^{-n^2 t} = \int_{S^1} e_{S^1}(t, \theta, \theta) \, d\theta = \int_{S^1} \sum_n e_{\mathbf{R}}(t, \theta, \theta + 2\pi n) \, d\theta$$

$$= \int_{-\pi}^{\pi} e_{\mathbf{R}}(t, x, x) \, dx + O(t^\infty),$$

where $O(t^\infty)$ denotes terms dying like $e^{-\frac{\alpha}{t}}$, for some $\alpha > 0$, as $t \to 0$. Thus as $t \to 0$, $\operatorname{Tr} e^{-t\Delta}$ on S^1 looks more and more like $2\pi/\sqrt{4\pi t}$. To be precise, we make the following definition:

Definition: *Given functions $A(t)$ and $B(t)$, we write $A(t) \sim B(t)$ if*

$$\lim_{t \to 0} \frac{A(t) - B(t)}{t^m} = 0,$$

for all $m \in \mathbf{R}^+$.

In this notation, we have shown the following result of Jacobi (ca. 1780):

Theorem 1.12

$$\sum_{n \in \mathbf{Z}} e^{-n^2 t} \sim \sqrt{\pi}\, t^{-\frac{1}{2}}.$$

In the course of this argument, we have seen that the short time behavior of the pointwise trace of the heat kernel $e_{S^1}(t, \theta, \theta)$ is the same as that of $e_{\mathbf{R}}(t, x, x)$, up to exponentially small factors. This is more or less plausible, if one argues that S^1 and $\mathbf{R}$ are locally isometric, and that for $t \approx 0$, all the information in the heat flow should be locally computable, as the heat has not "had enough time" to distribute itself around the manifold.

Moreover, if we take a circle of circumference ℓ, it is easy to check that Tr $e^{-t\Delta} \sim \ell/\sqrt{4\pi t}$, so the short time behavior of the heat operator recaptures the length, the only intrinsic geometric invariant of a circle (i.e. any two circles of the same length are isometric as metric spaces). In contrast, the long time behavior of the heat flow clearly differs on S^1 and $\mathbf{R}$, at least to the extent of distinguishing between a compact and a noncompact manifold.

These remarks are the simplest examples of quite general phenomena. We will see in this chapter that the long time behavior of the heat flow on functions and forms is determined by the topology of a compact manifold, and in Chapter 3 we will show that the short time behavior is controlled by the local differential geometry of the manifold. Moreover, comparing the long and short time behavior of the heat flow will lead in Chapter 4 to a proof of the Atiyah-Singer index theorem.

Finally, we should point out that it is unclear how closely the heat equation models actual heat flow. For example, the reader should check that (1.5) shows that $f(t, x)$ is a smooth function in x for any $t > 0$. Is it physically plausible that a discontinuous initial heat distribution in $L^2(\mathbf{R})$, such as a step function, should be immediately smoothed under heat flow, or is such an initial distribution physically implausible? Moreover, from (1.5) we see that for any $t > 0$, the heat distribution at x, namely $f(t, x)$, is affected by the initial distribution $f(y)$ at y, for y arbitrarily far from x. Thus, we say that heat flow has infinite propagation speed, in contrast to the solution of the wave equation. Is this physically plausible?

1.2   The Laplacian on a Riemannian Manifold

The goal of this section is to generalize the notion of the Laplacian on $L^2(S^1)$ to the Laplacian on L^2 functions on any manifold. To do this, we need to introduce a Riemannian metric on the manifold.

1.2.1   Riemannian Metrics

Given a smooth manifold, there is no natural way to define a generalization of the Laplacian on S^1 or on $\mathbf{R}$, without as additional data a "geometry" in the form of a Riemannian metric.

What is meant by geometry in this context? Certainly we should be able to measure lengths of curves on the manifold in order to do geometry. For a surface $M^2 \subset \mathbf{R}^3$, we can measure the length of a curve $\gamma : [0, 1] \to M$ by the usual formula

$$l(\gamma) = \int_0^1 |\gamma'(t)| dt.$$

Notice that the basic ingredient is the measure of the length of the tangent vector $\gamma'(t) \in T_{\gamma(t)}M$. We can also use this formula for any manifold embedded in $\mathbf{R}^n$, which covers the classical cases in algebraic geometry and analysis where manifolds appear as the zero set of constraint equations. However, not all manifolds arise in this way.

Exercise 5: *Show that the space of all positive definite inner products on $\mathbf{R}^n$ is in one-to-one correspondence with the set $M = GL(n; \mathbf{R})/O(n)$. Hint: Such an inner product is determined by a basis of $\mathbf{R}^n$ which is orthonormal with respect to the inner product. However, two bases which differ by an orthogonal transformation (with respect to this inner product) determine the same inner product.*

Because $O(n)$ is a closed subgroup of the Lie group $GL(n; \mathbf{R})$, M is a manifold [68]; this means we can smoothly parametrize the set of inner products. By Whitney's embedding theorem, M can be embedded in some Euclidean space. However, the embedding is not canonical, and so the geometry of M is not uniquely determined.

Thus in order to get a geometry on a manifold in the above sense, we need to introduce a method of measuring lengths of tangent vectors.

Definition: *A Riemannian manifold (M, g) is a smooth manifold M with a family of smoothly varying positive definite inner products $g = g_x$ on T_xM for each $x \in M$. The family g is called a Riemannian metric. Two Riemannian manifolds (M, g) and (N, h) are called* isometric *if there exists a smooth diffeomorphism $f : M \to N$ such that*

$$g_x(X, Y) = h_{f(x)}(f_* X, f_* Y)$$

for all $X, Y \in T_xM$, for all $x \in M$.

Note that since g_x is a bilinear form on T_xM, it is an element of $T_x^* M \otimes T_x^* M$. To say that g varies smoothly just means that g is a smooth section of the bundle $T^* M \otimes T^* M$. Given a Riemannian metric, we can set the length of a curve $\gamma : [0, 1] \to M$ to be

$$l(\gamma) = \int_0^1 g_x(\gamma'(t), \gamma'(t))^{\frac{1}{2}} dt.$$

Examples: (1) On $\mathbf{R}^n$, the standard Riemannian metric is given by the standard inner product $g_x(v, w) = v \cdot w$ for all $v, w \in T_x\mathbf{R}^n$, for all $x \in \mathbf{R}^n$. Of course, we call $\mathbf{R}^n$ with this Riemannian metric *Euclidean space*.

(2) If M is a submanifold of Euclidean space, then M has a natural Riemannian metric given by $g_x(v, w) = v \cdot w$. This so-called *induced metric* is the metric used in the classical theory of curves and surfaces in Euclidean three-space. By this same construction, a submanifold of a Riemannian manifold always inherits an induced Riemannian metric. Moreover, by the Whitney embedding theorem, every manifold possesses a Riemannian metric; this can also be shown by putting the metric of Example (1) on each chart in a cover of M and using a partition of unity argument.

(3) Let $M = GL(n; \mathbf{R})/O(n)$ as above.

Exercise 6: *(i) Show that for all $A \in GL(n; \mathbf{R})$, $T_A GL(n; \mathbf{R})$ is isomorphic to the space of $n \times n$ real valued matrices.*

(ii) Show that $g_A(X, Y) = \mathrm{Tr}\, XY^t$ is a Riemannian metric on $GL(n; \mathbf{R})$.

(iii) Consider $GL(n; \mathbf{R})$ as an open submanifold of $\mathbf{R}^{n^2}$. Show that the induced metric on $GL(n; \mathbf{R})$ equals g of (b).

(iv) $O(n)$ acts on $GL(n; \mathbf{R})$ by matrix multiplication. Show that this action preserves the Riemannian metric: if $O \in O(n)$ and $A \in GL(n; \mathbf{R})$, then $g_A(X, Y) = g_{OA}(O_*X, O_*Y)$. Conclude that the metric g descends to a metric h on the space of inner products $GL(n; \mathbf{R})/O(n)$. Hint: Let $\mathcal{O}_A$ be the orbit of a matrix $A \in GL(n; \mathbf{R})$ under the action of $O(n)$. Define $H_A \subset T_A GL(n; \mathbf{R})$ to be the orthogonal complement of $T_A \mathcal{O}_A$ in $T_A GL(n; \mathbf{R})$. Show that $H_{OA} = O_* H_A$. Let $[A]$ be the class of A in $M = GL(n; \mathbf{R})/O(n)$. Show that if $\pi : GL(n; \mathbf{R}) \to M$ is the projection, then $(\pi_*)_A : H_A \to T_{[A]} M$ is an isomorphism. Given $Z, W \in T_{[A]} M$, define

$$h(Z, W) = g((\pi_*)_B^{-1} Z, (\pi_*)_B^{-1} W)$$

for any choice of $B \in \mathcal{O}_A$. Show that h is well defined, since $O_*(\pi_*)_B^{-1} Z = (\pi_*)_A^{-1} Z$ if $OB = A$.

This last example shows that Riemannian metrics may arise naturally on manifolds which are not *a priori* embedded in $\mathbf{R}^n$. This raises the question of whether any compact Riemannian manifold can be embedded in Euclidean space in such a manner that the induced metric coincides with the original one. The Nash embedding theorem states that such an embedding is always possible; the proof is much more difficult than for the Whitney embedding theorem. Like the Whitney embedding theorem, the Nash embedding theorem is not of much practical help for questions in Riemannian geometry.

Exercise 7: *(i) Show that the space of metrics on a fixed manifold is a connected (in fact contractible) set. Hint: if g_0, g_1 are metrics, so is $tg_0 + (1 - t)g_1$ for $t \in [0, 1]$.*

(ii) (For those who know about infinite dimensional manifolds.) Show that the space of all Riemannian metrics on a fixed manifold M is an open submanifold of the space of symmetric contravariant two-tensors on M. Hint: a perturbation of a metric by a small symmetric two-tensor is still a Riemannian metric. What topology are we using when we say "open submanifold"?

To compute with a Riemannian metric, we must be able to analyze it in a local coordinate chart. If $v, w \in T_x M$ and $(x^1, x^2, \ldots, x^n)$ are coordinates near x, then there exist α^i, β^i such that

$$v = \sum \alpha^i \frac{\partial}{\partial x^i}, \quad w = \sum \beta^i \frac{\partial}{\partial x^i}.$$

We have

$$
\begin{aligned}
g_x(v, w) &= g_x\Big(\sum_i \alpha^i \partial_{x^i}, \sum_j \beta^j \partial_{x^j}\Big) \\
&= \sum_{i,j} \alpha^i \beta^j g_x(\partial_{x^i}, \partial_{x^j}),
\end{aligned}
$$

where $\partial_{x^i} = \partial/\partial x^i$. Thus, g_x is determined by the symmetric, positive definite matrix $(g_{ij}(x)) = (g_x(\partial_{x^i}, \partial_{x^j}))$. Note that while the metric g is defined on all of M, the $g_{ij}(x)$ are defined only in a coordinate chart, where we may write

$$g = \sum_{i,j} g_{ij} dx^i \otimes dx^j.$$

Exercise 8: *(i) Check that g is a Riemannian metric iff in any chart the functions $g_{ij}(x)$ are smooth functions for all i, j.*

(ii) Let $h_{ij} = g(\partial_{y^i}, \partial_{y^j})$ be the matrix of the metric g in another coordinate chart with coordinates $(y^1, \ldots, y^n)$. Show that on the overlap of the charts we have

$$g_{ij} = \sum_{k,\ell} \frac{\partial y^k}{\partial x^i} \frac{\partial y^\ell}{\partial x^j} h_{k\ell}.$$

Now that we can measure lengths of curves, it seems plausible to set the distance between any two points of M to be the length of the shortest path between them. However, such a path of shortest length need not exist; just consider the points $(1, 1)$ and $(-1, -1)$ in $\mathbf{R}^2 - \{(0, 0)\}$ with the metric induced from the Euclidean plane. We can avoid this problem by setting

$$d(x, y) = \inf_{\gamma \text{ piecewise } C^\infty} \{l(\gamma) : \gamma(0) = x, \gamma(1) = y\}.$$

Exercise 9: *Prove that (M,d) is a metric space. Hint: Use that γ need only be piecewise smooth to prove the triangle inequality. Don't forget to prove that $x \neq y$ implies $d(x, y) \neq 0$. Use the Weierstrass approximation theorem to conclude that d is a metric if we only use smooth curves.*

We will use this metric space structure associated to a Riemannian manifold in Chapter 3.

1.2.2 L^2 Spaces of Functions and Forms

At this point we can develop analysis on a manifold with metric. First of all, we want a Hilbert space of real valued functions on M, so it seems natural to define $L^2(M)$ by setting $\langle f, g \rangle = \int_M f(x) g(x)$. However, this is horribly wrong. We cannot integrate functions on a manifold; only n-forms transform correctly to give an integral over an n-manifold which is independent of coordinates. (This property in fact characterizes n-forms, a point often lost the first time around with differential forms.) It may be objected that we happily integrate functions on $\mathbf{R}^n$ and S^1, but this is possible only because we can identify a function f on S^1 with the one-form $f d\theta$ and a function f on $\mathbf{R}^n$ with $f dx^1 \wedge ... \wedge dx^n$. This identification occurs because these manifolds have special coordinates, which is precisely what general manifolds have not got.

From this point on we assume that the given manifold M is oriented and connected; this avoids technicalities later on. We are looking for an n-form $\alpha(x)$ such that $\langle f, g \rangle = \langle f, g \rangle_M = \int_M f(x) g(x) \alpha(x)$ defines a positive definite inner product; such an α is called a *volume form*. This terminology is motivated by noting that $\langle 1, 1 \rangle$ will then equal $\int_M \alpha$, and that for reasonable sets $A \subset \mathbf{R}^n$,

$$\text{vol}(A) = \int_A dx^1 \wedge ... \wedge dx^n = \langle 1, 1 \rangle_A.$$

We first compute what the volume of a Riemannian manifold should be. To make things simple, we'll just compute the volume of a coordinate chart; for the full volume, we would then use a partition of unity.

In a positively oriented coordinate neighborhood U around x with coordinates $(x^1, ..., x^n)$, pick a large number N of points p_j, and in each tangent space $T_{p_j} M$ form a box B_j with sides $(\Delta x^1) \frac{\partial}{\partial x^1}, (\Delta x^2) \frac{\partial}{\partial x^2}, ..., (\Delta x^n) \frac{\partial}{\partial x^n}$, for some small numbers Δx^i. Let $v_1, ..., v_n$ be a positively oriented orthonormal basis of $T_{p_j} M$. Then $\partial_{x^i} = \alpha_i^k v_k$ for some matrix $\alpha_i^k = \alpha_i^k(p_j)$. Here we are using the *Einstein summation convention*, which means that an index which appears as both a superscript and a subscript in an expression is summed over: e.g. $\alpha_i^k v_k \equiv \sum_k \alpha_i^k v_k$. With all this notation, we expect $\text{vol}(U)$ to be

$$
\begin{aligned}
\text{vol}(U) &= \lim_{\Delta x^i \to 0} \lim_{N \to \infty} \sum_{j=1}^{N} (\text{volume of } B_j) \\
&= \lim_{\Delta x^i \to 0} \lim_{N \to \infty} \sum_{j=1}^{N} \left(\text{volume of a box with } i^{\text{th}} \text{side } (\Delta x^i) \sum_{k=1}^{n} \alpha_i^k v_k \right) \\
&= \lim_{\Delta x^i \to 0} \lim_{N \to \infty} \sum_{j=1}^{N} \left((\Delta x^1)(\Delta x^2) ... (\Delta x^n) \det(\alpha_i^k) \right) \\
&= \int_U \det(\alpha_i^k) dx^1 \wedge ... \wedge dx^n.
\end{aligned}
$$

(We do not use summation convention for i in the second line.) For δ_{kl} the

Kronecker delta ($\delta_{kl} = 1$ if $k = l$, $\delta_{kl} = 0$ if $k \neq l$), we have

$$\begin{aligned}
g_{ij} &= \langle \partial_{x^i}, \partial_{x^j} \rangle = \langle \alpha_i^k v_k, \alpha_j^l v_l \rangle = \alpha_i^k \alpha_j^l \delta_{kl} \\
&= \alpha_i^k \alpha_j^k = (AA^t)_{ij}
\end{aligned}$$

where $A = (\alpha_i^j)$, and we denote $g(v, w)$ by $\langle v, w \rangle$. Thus

$$\det g = \det(AA^t) = (\det A)^2,$$

and so vol(U) should be $\int_U \sqrt{\det g}\ dx^1 \wedge \ldots \wedge dx^n$.

Exercise 10: *Show that $\sqrt{\det g}\ dx^1 \wedge \ldots \wedge dx^n$ is a well defined n-form on M^n: i.e. show that if $(y^1, \ldots, y^n)$ are also coordinates at x, then*

$$\sqrt{\det g(x)}\ dx^1 \wedge \ldots \wedge dx^n = \sqrt{\det g(y)}\ dy^1 \wedge \ldots \wedge dy^n.$$

Definition: *We define the* volume form *of a Riemannian metric to be the top dimensional form* dvol *which in local coordinates is given by*

$$\text{dvol} = \sqrt{\det g}\ dx^1 \wedge \ldots \wedge dx^n,$$

whenever $(\partial_{x^1}, \ldots, \partial_{x^n})$ is a positively oriented basis of $T_x M$. We set the volume *of (M, g) to be*

$$\text{vol}(M) = \int_M \text{dvol}(x).$$

Exercise 11: *(i) Show that* dvol *is bad notation: if M^n is compact and without boundary, then there is no $(n-1)$-form θ such that* dvol $= d\theta$. *A much harder exercise is to show that if M is noncompact, then there always exists such a θ.*

(ii) Let $f : \mathbf{R}^2 \to \mathbf{R}$ be a smooth function and let M be the graph of f with the metric induced from $\mathbf{R}^3$. Show that the volume element on M satisfies dvol $= \sqrt{1 + f_x^2 + f_y^2}\ dx \wedge dy$. *Compare this with the standard vector calculus definition of surface area.*

As an aside, notice that we usually define the length of a one-manifold, a curve, to be the limit of the sum of lengths of inscribed secants, while for surfaces the area is computed as the limit of sum of areas of boxes in tangent spaces. It turns out that we could equally well use tangent line segments in the curve case, but using inscribed planes in the surface case is problematic; very simple compact surfaces can have inscribed planes the sum of whose areas is arbitrarily large. Thus in our motivation we stick to summing up volumes of tangent boxes.

Exercise 12: *(i) Prove that a smooth n-manifold is orientable iff there exists a smooth nonzero section of $\Lambda^n T^* M$.*

(ii) Given such a nonzero section s, prove that there exists a Riemannian metric g on M with dvol(g) = s; you may have to switch the orientation of M to agree with the orientation determined by s. Hint: Pick any Riemannian metric g_1 on M. Find a function $f : M \to \mathbf{R}$ such that

$$\sqrt{\det(e^f g_1)}dx^1 \wedge \ldots \wedge dx^n = s$$

on M.

(iii) Show that $e^f g_1$ is conformal to g_1; i.e. define the angle between two intersecting curves, and show that the angle is the same when measured with respect to either $e^f g_1$ or g_1. Conversely, if two metrics g_1, g_2 always give the same angle measurements, show that there exists $f \in C^\infty(M)$ such that $g_1 = e^f g_2$.

(iv) Place a cylinder around the sphere S^2 so that the equator is the line of intersection. Map the sphere minus the two poles to the cylinder by radial projection from the center of the sphere. Show that this map, Mercator's projection, preserves angles, although it of course distorts distances.

Now we can define the Hilbert space $L^2(M, g)$ to be the completion of $C_c^\infty(M)$ (the smooth functions of compact support) with respect to the inner product $\langle f, g \rangle = \int_M f(x)g(x)\mathrm{dvol}$.

Exercise 13: Let M be a compact manifold with two Riemannian metrics g_1, g_2. Show that $L^2(M, g_1)$ and $L^2(M, g_2)$ are naturally isomorphic as Hilbert spaces. In contrast, if M is noncompact, show that there always exist metrics g_1, g_2 such that these two Hilbert spaces are not naturally isomorphic, in the sense that there exist smooth functions in one Hilbert space but not in the other.

We'll also need spaces of L^2 k-forms. For one-forms this is fairly easy. Recall that if V is a finite dimensional vector space with an inner product $\langle \, , \, \rangle$, then the dual vector space V^* is naturally isomorphic to V under the map $\alpha : V \to V^*$, where $\alpha(v) = v^*$ satisfies $v^*(w) = \langle v, w \rangle$ for $v, w \in V$. V^* inherits an inner product, also denoted $\langle \, , \, \rangle$, given by $\langle v^*, w^* \rangle = \langle v, w \rangle$. In particular, a Riemannian metric g on a manifold produces an inner product, also denoted g, on each cotangent space $T_x^* M$ under the isomorphism $\alpha = \alpha_{g,x}$ constructed above. We'll also use α to denote the bundle isomorphism $\alpha : TM \to T^*M$ induced by $\alpha_{g,x}$ in each fiber.

Exercise 14: In local coordinates, set $g^{ij} = g(dx^i, dx^j)$. Then $g^{ik}g_{kj} = \delta_j^i$, where δ_j^i is the Kronecker delta – i.e. $(g^{ij}) = (g_{ij})^{-1}$.

We now set $L^2\Lambda^1 T^*(M, g)$ (or just $L^2\Lambda^1 T^*M$ for short) to be the completion of $C_c^\infty T^*M$ with respect to the global inner product

$$\langle \omega, \eta \rangle = \int_M g(\omega, \eta) \, \mathrm{dvol}(x).$$

Similarly, the inner product g induces an inner product g on each tensor product $T_x M \otimes \ldots \otimes T_x M$ and hence on each exterior power $\Lambda^k T_x^* M$, and hence a global inner product

$$\langle \alpha, \beta \rangle = \int_M g(\alpha, \beta) \, \mathrm{dvol}$$

for $\alpha, \beta \in C_0^\infty \Lambda^k T^* M$. The completion is denoted by $L^2 \Lambda^k T^* M$.

Exercise 15: *Compute* $g(dx^i \wedge dx^j, dx^k \wedge dx^l)$ *in terms of* (g^{ij}).

1.2.3   The Laplacian on Functions

Now that we have the Hilbert spaces in place, it is time to define the Laplacian $\Delta : L^2(M, g) \to L^2(M, g)$. (Keep in mind that Δ will only be defined on a dense subset of $L^2(M)$.) Of course, we want the Laplacian to agree with (minus) the standard Laplacian $-(\frac{\partial^2}{\partial (x^1)^2} + \ldots + \frac{\partial^2}{\partial (x^n)^2})$ on $\mathbf{R}^n$. However, this expression depends on the standard coordinates for $\mathbf{R}^n$, and we need a coordinate free expression for our generalization. This is provided by the classical equation

$$-(\frac{\partial^2}{\partial (x^1)^2} + \ldots + \frac{\partial^2}{\partial (x^n)^2}) = -\mathrm{div} \circ \nabla.$$

Now do we have the operators $\nabla : C^\infty(M) \to TM$ and $\mathrm{div} : TM \to C^\infty(M)$ on a general manifold? The answer is no, but on a Riemannian manifold we may set ∇ to be the composition

$$C^\infty(M) \overset{d}{\to} \Lambda^1 T^* M \underset{\cong}{\overset{\alpha_g^{-1}}{\to}} TM.$$

It is easy to check that this produces the ordinary gradient in Euclidean space.

Exercise 16: *Show that in local coordinates, we have*

$$\nabla f = g^{ij} \partial_i f \partial_j,$$

where $\partial_j = \partial_{x^j} = \frac{\partial}{\partial x^j}$.

As for div, integration by parts applied to $f \in C_c^\infty(\mathbf{R}^n)$ gives

$$-\int_{\mathbf{R}^n} \partial_i X^i \cdot f = \int_{\mathbf{R}^n} \partial_i f \cdot X^i,$$

for functions X^i, which shows that the divergence $\partial_i X^i$ of a vector field $X = X^i \partial_i$ on $\mathbf{R}^n$ is characterized by the equation

$$\langle -\mathrm{div} X, f \rangle = \langle X, \nabla f \rangle, \tag{1.13}$$

where the inner products are the global inner products on functions and vector fields induced by the standard dot product. In other words, $-\mathrm{div}$ is the (formal) adjoint to ∇. We would like to use (1.13) on a Riemannian manifold to define div X, since we already know what ∇f and the inner products in (1.13) mean on a Riemannian manifold. (The inner product on the right hand side is the global inner product on TM defined just as for T^*M.)

Exercise 17: *Why is the following attempt to define* div X *from (1.13) not valid? Recall that the dual space to a Hilbert space H is canonically isomorphic to H via the map α above: i.e. for all functionals $\lambda \in H^*$, there exists a unique $v \in H$ such that $\lambda = \alpha(v) = \langle v, \cdot \rangle$. For a fixed vector field X on a Riemannian manifold, the right hand side of (1.13) defines a linear functional λ_X on $L^2(M)$, and so we can define $-$div X to be the unique function $v \in L^2(M)$ satisfying $\langle v, f \rangle = \lambda_X(f)$, which is just (1.13).*

Assume for the moment that an operator div X satisfying (1.13) exists. What must this operator look like in local coordinates? Let U be a coordinate patch on M; we won't distinguish in the notation between integration over U and integration over the coordinate chart image of U in $\mathbf{R}^n$. For any function $f \in C_c^\infty(U)$ and vector field $X = X^i \partial_i \in TM$, we have

$$
\begin{aligned}
\langle X, \nabla f \rangle &= \int_M \langle X, \nabla f \rangle \, d\mathrm{vol} \\
&= \int_U \langle X^i \partial_i, g^{kj} \partial_k f \partial_j \rangle \, d\mathrm{vol} \\
&= \int_U X^i (\partial_k f) g^{kj} g_{ij} \sqrt{\det g} \, dx^1 \ldots dx^n \\
&= \int_U X^i (\partial_i f) \sqrt{\det g} \, dx_1 \ldots dx^n \\
&= -\int_U \frac{1}{\sqrt{\det g}} f \cdot \partial_i (X^i \sqrt{\det g}) \sqrt{\det g} \, dx_1 \ldots dx^n \\
&= \langle f, -\frac{1}{\sqrt{\det g}} \partial_i (X^i \sqrt{\det g}) \rangle.
\end{aligned}
$$

Thus, if div X exists, it must satisfy

$$
\mathrm{div}\, X = \frac{1}{\sqrt{\det g}} \partial_i (X^i \sqrt{\det g}).
$$

Assuming this expression is independent of choice of coordinates, we can then define the Laplacian on functions to be $\Delta = -\mathrm{div} \circ \nabla$, a second order differential operator. In local coordinates, we get

$$
\begin{aligned}
\Delta f &= -\frac{1}{\sqrt{\det g}} \partial_j (g^{ij} \sqrt{\det g} \, \partial_i f) \\
&= -g^{ij} \partial_j \partial_i f + \text{(lower order terms).} \quad (1.14)
\end{aligned}
$$

Note that this reduces to the usual expression for the Laplacian on $\mathbf{R}^n$. The last expression shows that not only is the Laplacian determined by the Riemannian metric, but the Laplacian also determines the metric. (By evaluating Δ on a function which is locally $x^i x^j$, we can recover g^{ij} and hence g_{ij}.) Thus we expect the spectral theory of the Laplacian to be intimately connected with the geometry of (M, g).

Exercise 18: *Show that* div X *is well defined: i.e. given another set of coordinates* $(y^1, ..., y^n)$ *on* U, *write* $X = X^i \partial_{x^i} = Y^j \partial_{y^j}$ *and show that*

$$\frac{1}{\sqrt{\det g}} \partial_{x^i} (X^i \sqrt{\det g}) = \frac{1}{\sqrt{\det g}} \partial_{y^j} (Y^j \sqrt{\det g}).$$

By this exercise, the Laplacian is well defined. However, it is much nicer to construct the Laplacian in a coordinate free manner from the start, so that there is no need to do local calculations as a consistency check. Before we do this, we rewrite (1.13) in terms of one-forms, as this is more convenient for later purposes. Because α is trivially an isometry, it is easy to check that at each point of M, we have $g(\alpha(X), df) = g(X, \nabla f)$ for any tangent vector X and any function f. Set $\delta : \Lambda^1 T^* M \to C^\infty(M)$ by $\delta(\omega) = -\text{div}(\alpha^{-1}(\omega))$, so that δ is "the same" as div up to the isomorphism $\alpha : TM \to T^* M$. Then δ, which exists by the last exercise, is characterized by the equation

$$\langle \delta\omega, f \rangle = \langle \omega, df \rangle, \tag{1.15}$$

for all $\omega \in C_c^\infty \Lambda^1$, $f \in C_c^\infty$.

Exercise 19: *Check that* δ *is given by*

$$\delta(\omega) = -\frac{1}{\sqrt{\det g}} \partial_i (g^{ij} \sqrt{\det g}\ \omega_i),$$

where $\omega = \omega_i dx^i$, *and that this expression is independent of choice of local coordinates.*

As a second (coordinate dependent) definition, we define the Laplacian by $\Delta = \delta d$. This is the same as our first definition, since $\delta df = (-\text{div } \alpha^{-1})(\alpha \nabla f) = -\text{div } \nabla f$. The reader might prefer to check in local coordinates that these two definitions agree.

We now introduce the *Hodge star operator*, which is a pointwise isometry $* = *_x : \Lambda^k T_x^* M \to \Lambda^{n-k} T_x^* M$. Choose a positively oriented orthonormal basis $\{\theta^1, ..., \theta^n\}$ of $T_x^* M$. Since $*$ is a linear transformation, we just need to define $*$ on a basis element $\theta^{i_1} \wedge ... \wedge \theta^{i_k}$ ($i_1 < ... < i_k$) of $\Lambda_x^k T^* M$. Note that

$$\begin{aligned} \text{dvol}(x) &= \sqrt{\det(\langle \theta^i, \theta^j \rangle)}\ \theta^1 \wedge ... \wedge \theta^n \\ &= \theta^1 \wedge ... \wedge \theta^n. \end{aligned}$$

Definition: $*(\theta^{i_1} \wedge ... \wedge \theta^{i_k}) = \theta^{j_1} \wedge ... \wedge \theta^{j_{n-k}}$ *where* $\theta^{i_1} \wedge ... \wedge \theta^{i_k} \wedge \theta^{j_1} \wedge ... \wedge \theta^{j_{n-k}} = \text{dvol}(x)$.

This definition forces

$$\{j_1, \ldots, j_{n-k}\} = \{1, 2, \ldots, n\} - \{i_1, \ldots, i_k\},$$

and we always have

$$\theta^{i_1} \wedge \ldots \wedge \theta^{i_k} \wedge *(\theta^{i_1} \wedge \ldots \wedge \theta^{i_k}) = \text{dvol}.$$

In particular, $*1 = \text{dvol}$ and $*\text{dvol} = 1$. The reader should check that the definition above is independent of the choice of orthonormal basis, and that $*^2 = (-1)^{k(n-k)}\text{Id}$ on $\Lambda^k T_x^* M$, where $\dim M = n$.

Exercise 20: *Show that the equation*

$$\langle\langle \omega, \eta \rangle\rangle = *(\omega \wedge *\eta)$$

defines a positive definite inner product $\langle\langle \ , \ \rangle\rangle$ on each exterior power $\Lambda^k T_x^* M$.

Since the Riemannian metric g induces the inner product denoted $\langle \ , \ \rangle$ on $\Lambda^k T_x^* M$, we now have two inner products associated to the metric. Fortunately, these two are the same, as we now show:

Claim: For $\omega, \eta \in \Lambda^k T_x^* M$,

$$\langle \omega, \eta \rangle_x \, \text{dvol} = \omega \wedge *\eta.$$

PROOF: Both sides of the equation are linear in ω and η, so we may assume

$$\omega = \theta^{i_1} \wedge \ldots \wedge \theta^{i_k}, \quad \eta = \theta^{l_1} \wedge \ldots \wedge \theta^{l_k}.$$

Then

$$
\begin{aligned}
\langle \omega, \eta \rangle \, \text{dvol} &= \langle \theta^{i_1} \wedge \ldots \wedge \theta^{i_k}, \theta^{l_1} \wedge \ldots \wedge \theta^{l_k} \rangle \, \text{dvol} \\
&= \pm \delta^{\{i_1, \ldots, i_j\}}_{\{l_1, \ldots, l_j\}} \, \text{dvol}
\end{aligned}
$$

(the last step follows from $\langle \theta^i, \theta^j \rangle = \delta^i_j$). We also have $\omega \wedge *\eta = (\theta^{i_1} \wedge \ldots \wedge \theta^{i_k}) \wedge *(\theta^{l_1} \wedge \ldots \wedge \theta^{l_k}) = \pm \delta^{\{i_1, \ldots, i_j\}}_{\{l_1, \ldots, l_j\}} \, \text{dvol}$. We leave it to the reader to check that the plus/minus sign is the same in both cases.

Exercise 21: *(i) Use this claim to show that the Hodge star is an isometry.*
(ii) Let $\omega = \omega_i dx^i$ be a one-form written in local coordinates. Compute $*\omega$ in local coordinates. Hint: Write $dx^i = b^i_j \theta^j$ for some invertible matrix $B = (b^i_j)$. Note that $BB^t = (g^{ij})$. Then

$$*\omega = \sum_j c_j dx^1 \wedge \ldots \wedge dx^{j-1} \wedge dx^{j+1} \wedge \ldots \wedge dx^n,$$

where c_j is a complicated expression in the entries of B. Rewrite this expression in terms of the g^{ij} and $\det g$.

Thus on k-forms we may express the global inner product (also known as the Hodge inner product) in two ways:

$$\langle \omega, \eta \rangle = \int_M g(\omega, \eta) \ \mathrm{dvol} = \int_M \omega \wedge *\eta.$$

For the exterior derivative $d : \Lambda^k \to \Lambda^{k+1}$, we have

$$
\begin{aligned}
\langle d\omega, \alpha \rangle &= \int_M d\omega \wedge *\alpha \\
&= \int_M d(\omega \wedge *\alpha) - (-1)^k \int \omega \wedge d*\alpha \\
&= (-1)^{k+1} \int \omega \wedge d*\alpha \\
&= (-1)^{k+1}(-1)^{(n-k)(n-(n-k))} \int \omega \wedge **d*\alpha \\
&= (-1)^{nk+1} \int \omega \wedge *(*d*\alpha) \\
&= (-1)^{nk+1} \langle \omega, *d*\alpha \rangle,
\end{aligned}
$$

where we have used Stokes' theorem in the third line. Thus the adjoint δ^{k+1} of d on k-forms is given by $\delta^{k+1} = (-1)^{nk+1} *d*$, and in particular, $\delta = \delta^1 = -*d*$.

Exercise 22: *Compute that $-*d*$ on one-forms has the same local expression as in Exercise 19.*

Thus the Hodge star construction and Stokes' theorem (which is the coordinate free version of integration by parts on a manifold) lead to a coordinate free description of the adjoint of exterior differentiation. We are now entitled to define the Laplacian:

Definition: *The* Laplacian on functions *on a Riemannian manifold is given by*

$$\Delta f = \delta df = -*d*df.$$

It is interesting to review classical vector analysis in $\mathbf{R}^3$ in light of the techniques of this section. The modern version of Stokes' theorem, as taught in courses on the differential topology of manifolds, involves no Riemannian metric. However, to reinterpret the classical Green's, Stokes' and divergence theorems as special cases of the modern Stokes' theorem requires the use of the standard dot product metric on $\mathbf{R}^3$; see Exercise 11(ii) and [63]. Moreover, vector calculus in $\mathbf{R}^3$ is full of unstated identifications of tangent spaces and cotangent spaces via the α map and between one-forms and two-forms via the Hodge star operator. For example, it is easy to check that the classical curl operator taking vector fields to vector fields is given by

$$\text{curl } X = \alpha^{-1} * d\alpha X.$$

1.3   Hodge Theory for Functions and Forms

In this section we will prove the Hodge theorem for compact connected oriented Riemannian manifolds, which states that, just as on S^1, there is an orthonormal basis of $L^2(M, g)$ such that the Laplacian diagonalizes with respect to this basis:

$$\Delta \sim \begin{pmatrix} \lambda_1 & & & \\ & \lambda_2 & & \\ & & \ddots & \end{pmatrix}.$$

We will also prove the corresponding theorem for the Laplacian acting on k-forms.

1.3.1   Analytic Preliminaries

The proof we give, which is modeled on the heat equation proof of Milgram and Rosenbloom [44], depends on some standard Sobolev space analysis in Euclidean space, suitably generalized to compact manifolds.

We first recall the definition of the Sobolev spaces of complex valued functions on a set $\Omega \subset \mathbf{R}^n$, where Ω is open with compact closure $\overline{\Omega}$. The sth Sobolev space is a Banach space whose norm measures the L^2 norm of a function and its first s derivatives. (Recall that the usual Banach space norms on functions measure the sup norm of the first s derivatives.) In other words, the Sobolev space $H_s(\Omega)$, for $s \in \{0, 1, 2, \ldots\}$, is the completion of $C_c^\infty(\Omega)$ with respect to the norm

$$\|f\|_s = \left(\sum_{|\alpha| \le s} \|D^\alpha f\|_2^2 \right)^{\frac{1}{2}},$$

where $\alpha = (\alpha_1, \ldots, \alpha_n)$, $\alpha_j \in \mathbf{Z}$ is a multi-index with $|\alpha| = \sum_j \alpha_j$ and

$$D^\alpha = (-i)^{|\alpha|} \frac{\partial^{\alpha_1 + \ldots + \alpha_k}}{\partial_{x^1}^{\alpha_1} \ldots \partial_{x^n}^{\alpha_n}}. \tag{1.16}$$

Note that $H_0(\Omega) = L^2(\Omega)$. Of course, we can also define Sobolev spaces of real valued functions (omitting the $(-i)^{|\alpha|}$ in D^α), and the analytic results below remain valid.

We recall the basic properties of the Fourier transform on $\mathbf{R}^n$, defined by

$$\hat{u}(\xi) = \frac{1}{(2\pi)^{n/2}} \int_{\mathbf{R}^n} e^{-ix\cdot\xi} u(x) \, dx.$$

We will also use the notation $(f)\check{}$ for complicated functions f. Define the convolution of functions on $\mathbf{R}^n$ by

$$(u * v)(x) = \int_{\mathbf{R}^n} u(x - y)v(y) \; dy.$$

Here and from now on we use the measure dy which is $(2\pi)^{-n/2}$ times Lebesque measure $dy^1 \ldots dy^n$ to avoid a plague of 2π factors. The proof of the following lemma can be found in many texts.

Lemma 1.17 *(i) The Fourier transform is an isometry on $C_c^\infty(\mathbf{R}^n)$ in the L^2 norm, and so extends to an isometry of $L^2(\mathbf{R}^n)$.*
(ii) $\widehat{u * v} = \hat{u} \cdot \hat{v}, \; \widehat{uv} = \hat{u} * \hat{v}$.
(iii) $u(x) = \int_{\mathbf{R}^n} e^{ix \cdot \xi} \hat{u}(\xi) \; d\xi$.
(iv) $\widehat{D^\alpha u}(\xi) = \xi^\alpha \hat{u}(\xi)$. (Here $\xi^\alpha = \xi_1^{\alpha_1} \cdot \ldots \cdot \xi_n^{\alpha_n}$ for $\xi = (\xi_1, \ldots, \xi_n)$.)
(v) $\widehat{x^\alpha u}(\xi) = D^\alpha \hat{u}(\xi)$.

The proofs of the basic analytic results depend upon an alternative definition of the Sobolev spaces. Given two norms $\| \cdot \|^{(1)}$ and $\| \cdot \|^{(2)}$ on a vector space, we write $\| \cdot \|^{(1)} \approx \| \cdot \|^{(2)}$ if there exist positive constants C_1, C_2 such that $C_1 \|f\|^{(1)} \le \|f\|^{(2)} \le C_2 \|f\|^{(1)}$ for all f in the vector space.

Lemma 1.18 *For $f \in C_c^\infty(\Omega)$, we have*

$$\|f\|_s \approx \left(\int_{\mathbf{R}^n} |\hat{f}(\xi)|^2 (1 + |\xi|^2)^s d\xi \right)^{\frac{1}{2}}. \tag{1.19}$$

Since the right hand side of (1.19) is defined for all $s \in \mathbf{R}$, we can define Sobolev spaces $H_s = H_s(\Omega)$ for any real s to be the completion of $C_c^\infty(\Omega)$ with respect to

$$\|f\|_s = \left(\int_{\mathbf{R}^n} |\hat{f}(\xi)|^2 (1 + |\xi|^2)^s d\xi \right)^{\frac{1}{2}}.$$

This extends the previous definition, as the Hilbert space completions of a pre-Hilbert space with respect to two equivalent norms are naturally isomorphic (check this). Using the fact that $(1 + |\xi|^2)^s > (1 + |\xi|^2)^t$ for $s > t$, it is easy to see that if $s > t > 0 > r$, we have continuous inclusions

$$H_s \hookrightarrow H_t \hookrightarrow H_0 = L^2 \hookrightarrow H_r;$$

it should be trivial to check this for $s, t \in \mathbf{Z}^+$ using the first definition of Sobolev spaces.

PROOF OF LEMMA 1.18. By the previous lemma, we have

$$\begin{aligned}
\|f\|_s^2 &= \sum_{|\alpha| \le s} \|(D^\alpha f)\check{}\|_2^2 = \sum_{|\alpha| \le s} \int_{\mathbf{R}^n} |\xi^\alpha \hat{f}(\xi)|^2 \; d\xi \\
&= \int_{\mathbf{R}^n} \left(\sum_{|\alpha| < s} |\xi^\alpha|^2 \right) |\hat{f}(\xi)|^2 \; d\xi.
\end{aligned}$$

There exist constants C_1, C_2 such that

$$C_1(1 + |\xi|^2)^s \leq (\sum_{|\alpha| \leq s} |\xi|^\alpha|^2) \leq C_2(1 + |\xi|^2)^s,$$

since all three terms are polynomials in ξ of the same degree. The lemma follows.

We need two important theorems about Sobolev spaces. First, note that $f \in H_k(\Omega)$ if $f \in C^k(\overline{\Omega})$. The Sobolev embedding theorem is a partial converse to this.

Theorem 1.20 (Sobolev Embedding Theorem)

$$f \in H_k(\Omega) \Longrightarrow f \in C^s(\overline{\Omega}), \ \forall s < k - \frac{n}{2}.$$

Corollary 1.21 $f \in \bigcap_{k \in \mathbf{R}} H_k(\Omega) \Longleftrightarrow f \in C^\infty(\overline{\Omega})$.

PROOF OF THEOREM 1.20. We first do the case $s = 0$ – i.e. if $k > n/2$, then $f \in H_k \Rightarrow f \in C^0$. We have

$$
\begin{aligned}
|f(x)| &= \left| \int_{\mathbf{R}^n} e^{ix \cdot \xi} \hat{f}(\xi) \, d\xi \right| \\
&= \left| \int_{\mathbf{R}^n} e^{ix \cdot \xi} (1 + |\xi|^2)^{-k/2}(1 + |\xi|^2)^{k/2} \hat{f}(\xi) \, d\xi \right| \\
&\leq \left(\int_{\mathbf{R}^n} (1 + |\xi|^2)^{-k} \, d\xi \right)^{1/2} \left(\int_{\mathbf{R}^n} |\hat{f}(\xi)|^2 (1 + |\xi|^2)^k \, d\xi \right)^{1/2}
\end{aligned}
$$

by Cauchy-Schwarz. Now $k > n/2$ implies that $\int_{\mathbf{R}^n} (1 + |\xi|^2)^{-k} \, d\xi = C$ is finite, so

$$|f(x)| \leq C^{1/2} \|f\|_k.$$

Any $f \in H_k$ is the H_k-limit of compactly supported smooth functions f_i. The last equation implies that $f_i \to f$ uniformly, and so f is continuous.

For the general case, fix s and choose $f \in H_k$ with $k > s + (n/2)$. For any multi-index α with $|\alpha| \leq s$, we have

$$
\begin{aligned}
\|D^\alpha f\|_{k-|\alpha|}^2 &= \int_{\mathbf{R}^n} |(D^\alpha f)\hat{\ }|^2 (1 + |\xi|^2)^{k-|\alpha|} \, d\xi \\
&= \int_{\mathbf{R}^n} (|\xi^\alpha| \, |\hat{f}(\xi)|)^2 (1 + |\xi|^2)^{k-|\alpha|} \, d\xi \\
&\leq C \|f\|_k,
\end{aligned}
$$

since as above for some constant C, we have

$$|\xi^\alpha|^2 (1 + |\xi|^2)^{k-|\alpha|} \leq C(1 + |\xi|^2)^k.$$

This implies that $D^\alpha : H_k \to H_{k-|\alpha|}$ is continuous. By the first part of the proof, we conclude that $D^\alpha f \in C^0$ for all $|\alpha| \leq s$, and so $f \in C^s$.

Recall that a map of one Banach space to another is called *compact* if the image of every bounded sequence contains a convergent sub-sequence. In particular, the identity map is never compact on an infinite dimensional Banach space.

Theorem 1.22 (Rellich-Kondarachov Compactness Theorem) *If $t > s$, then the inclusion $H_t(\Omega) \to H_s(\Omega)$ is compact.*

PROOF. Let $\{f_n\} \subset H_t$ have $\|f_n\|_t \leq 1$. We first show that a sub-sequence of $\{f_n\}$ converges uniformly on compact subsets of $\mathbf{R}^n$. Choose a smooth compactly supported function ϕ with $\phi \equiv 1$ on Ω. Letting $D_j = (-i)\partial/\partial\xi_j$, we have

$$
\begin{aligned}
|D_j \hat{f}_n(\xi)| &= |D_j(\phi \cdot f_n)\check{}\,(\xi)| = |D_j(\hat{\phi} * \hat{f}_n)(\xi)| \\
&= \left| \int_{\mathbf{R}^n} D_j\hat{\phi}(\xi - \eta)\hat{f}_n(\eta)\, d\eta \right| \\
&\leq \left(\int_{\mathbf{R}^n} |D_j\hat{\phi}(\xi - \eta)|^2(1 + |\eta|^2)^{-t}\, d\eta \right)^{1/2} \\
&\qquad \times \left(\int_{\mathbf{R}^n} |\hat{f}_n(\eta)|^2(1 + |\eta|^2)^t\, d\eta \right)^{1/2} \\
&= C(\xi)\|f_n\|_t \leq C(\xi),
\end{aligned}
$$

where we have used Cauchy-Schwarz and set

$$
C(\xi) \equiv \left(\int_{\mathbf{R}^n} |D_j\hat{\phi}(\xi - \eta)|^2(1 + |\eta|^2)^{-t}\, d\eta \right)^{1/2}.
$$

Since $C(\xi)$ is a continuous function, it is bounded on compact ξ-sets.

This shows that each $D_j\hat{f}_n$ is uniformly bounded on compact ξ-sets, and a similar argument shows the same for $\hat{f}_n$. The Arzela-Ascoli theorem now guarantees the existence of a sub-sequence of $\{\hat{f}_n\}$, which we still call $\{\hat{f}_n\}$, converging uniformly on compact ξ-sets.

Since H_s is complete, the proof will be finished once we show that $\{f_n\}$ is a Cauchy sequence. For fixed $r > 0$,

$$
\begin{aligned}
\|f_n - f_m\|_s^2 &= \int_{\mathbf{R}^n} |\hat{f}_n(\xi) - \hat{f}_m(\xi)|^2(1 + |\xi|^2)^s\, d\xi \\
&= \int_{|\xi| \leq r} |\hat{f}_n(\xi) - \hat{f}_m(\xi)|^2(1 + |\xi|^2)^s\, d\xi \\
&\qquad + \int_{|\xi| > r} |\hat{f}_n(\xi) - \hat{f}_m(\xi)|^2(1 + |\xi|^2)^s\, d\xi.
\end{aligned}
$$

Since $t > s$, we have $(1 + |\xi|^2)^s (1 + r^2)^{t-s} \leq (1 + |\xi|^2)^t$ for $|\xi| > r$. This gives

$$\int_{|\xi|>r} |\hat{f}_n(\xi) - \hat{f}_m(\xi)|^2 (1 + |\xi|^2)^s \, d\xi$$

$$\leq \frac{1}{(1+r^2)^{t-s}} \int_{|\xi|>r} |\hat{f}_n(\xi) - \hat{f}_m(\xi)|^2 (1 + |\xi|^2)^t \, d\xi$$

$$= \frac{1}{(1+r^2)^{t-s}} \|f_n - f_m\|_t^2$$

$$\leq \frac{4}{(1+r^2)^{t-s}}.$$

For r sufficiently large, this term is as small as desired. Moreover, for fixed r and m, n sufficiently large,

$$\int_{|\xi|\leq r} |\hat{f}_n(\xi) - \hat{f}_m(\xi)|^2 (1 + |\xi|^2)^s \, d\xi$$

is also as small as desired, since $\{\hat{f}_n\}$ converges uniformly on the compact set $|\xi| \leq r$. This shows that $\{f_n\}$ is a Cauchy sequence in H_s.

It is straightforward, although a little messy, to define Sobolev spaces on manifolds. Let $\{U_i \subset \mathbf{R}^n, \varphi_i\}$, where $\varphi_i : U_i \to M$, be a locally finite coordinate cover of M, with $\overline{U_i}$ compact in $\mathbf{R}^n$. Take a partition of unity $\{\rho_i\}$ subordinate to U_i. Set $H_s(M)$ to be the completion of $C_c^\infty(M)$ with respect to

$$\|f\|_s = (\sum_i \|(\rho_i \cdot f) \circ \varphi_i\|_s^2)^{\frac{1}{2}}.$$

Exercise 23: *(i) Let $\{V_j, \psi_j\}$ be another locally finite cover with subordinate partition of unity $\{\mu_j\}$. Show that*

$$\|f\|_s^{U_i,\varphi_i,\rho_i} \approx \|f\|_s^{V_j,\psi_j,\mu_j} \tag{1.23}$$

if M is compact. If M is noncompact, show that there always exist covers $\{U_i\}, \{V_j\}$ such that (1.23) fails.

(ii) For a compact manifold M, show that $H_0(M)$ is naturally isomorphic to $L^2(M, g)$, for any choice of Riemannian metric g on M, by showing that $\|f\|_0 \approx \langle f, f \rangle_g^{1/2}$.

It is not too surprising that the embedding theorem holds on any manifold, once Sobolev spaces are defined correctly, since differentiability properties of functions are local properties; i.e. one checks them locally, and locally a manifold is indistinguishable from our Ω. In contrast, the compactness theorem always fails if $\overline{\Omega}$ is noncompact. Just take a sequence f_i with $\|f_i\|_1 = 1$ and with the supports of the f_i pairwise disjoint; then $\{f_i\}$ has no convergent sub-sequence in $L^2(\overline{\Omega})$. This example carries over to any noncompact manifold. Because the compactness theorem fails for noncompact spaces, the Hodge theorem as stated is not valid for noncompact manifolds. This "explains" why the Laplacian on **R** does not have an eigenfunction decomposition as on S^1. In any case, the embedding theorem and compactness theorem are valid for compact manifolds.

Exercise 24: *Prove the compactness theorem for compact manifolds, assuming the compactness theorem for $\overline{\Omega}$ as above. Hint: Take a sequence $\{f_j\} \subset H_t(M)$ with $\|f_j\|_t$ bounded. Show that in the ith coordinate chart U_i, $\{\rho_i f_j \varphi_i\}$ satisfies the hypothesis of the compactness theorem in $\mathbf{R}^n$, and so has a convergent subsequence in $H_s(U_i)$. Now use the compactness of M.*

We will also need the existence of a *heat kernel* analogous to that on S^1. Namely, we claim that there exists $e(t, x, y) \in C^\infty(\mathbf{R}^+ \times M \times M)$ such that

$$\left.\begin{aligned} (\partial_t + \Delta_x)e(t, x, y) &= 0, \\[2ex] \lim_{t \to 0} \int_M e(t, x, y)f(y)dy &= f(x). \end{aligned}\right\} \tag{1.24}$$

Here Δ_x denotes the Laplacian acting in the x variable. The existence of $e(t, x, y)$ will occupy Chapter 3.

Exercise 25: *Show that there is at most one function $e(t, x, y)$ on M satisfying (1.24). Moreover, we have $e(t, x, y) = e(t, y, x)$. Hint: imitate the proof of Lemma 1.9.*

If we define the *heat operator* $e^{-t\Delta}$ by $e^{-t\Delta}f(x) = \int_M e(t, x, y)f(y)\,\mathrm{dvol}(y)$, then $e^{-t\Delta}f(x)$ solves the heat equation with initial condition (temperature distribution) $f(x) \in L^2(M)$: setting $f(t, x) = e^{-t\Delta}f(x)$, we have

$$\begin{aligned} (\partial_t + \Delta_x)f(t, x) &= 0, \\ f(0, x) &= f(x). \end{aligned}$$

Here $f(0, x) \equiv \lim_{t \to 0} f(t, x)$.

Exercise 26: *Use (1.24) to verify these last two equations.*

1.3.2 The Heat Equation Proof of the Hodge Theorem for Functions

We now assume that M is compact. We use the abbreviation $dy = \mathrm{dvol}(y)$ to denote the variable of integration. In a coordinate chart, by abuse of notation we may write

$$D_x^\alpha e^{-t\Delta}f(x) = \int_M D_x^\alpha e(t, x, y)f(y)dy.$$

(How can this be written correctly?) Since $e(t, x, y)$ is smooth in x, we see that for $f \in L^2(M)$, $e^{-t\Delta}f \in C^k(M)$ for all t, k, and so $e^{-t\Delta}f \in H_s(M)$ for all t, s. Moreover, we claim that if $f_i \to 0$ in $L^2(M)$, then $e^{-t\Delta}f_i \to 0$ in H_1.

Exercise 27: *Prove the last statement. Hint: If for fixed t, $e(t, x, y)$ is a smooth function on the product of the closure of a domain $\bar{\Omega} \times \bar{\Omega}$, then $f_i \to 0$ in $L^2(\Omega)$ implies*

$$\partial_j \int_\Omega e(t, x, y)f_i(y)dy = \int_\Omega \partial_j e(t, x, y)f_i(y)dy \to 0$$

uniformly on $\bar{\Omega}$. Conclude that $\|\partial_j e^{-t\Delta}f_i\|_2 \to 0$; similarly $\|e^{-t\Delta}f_i\|_2 \to 0$. Now use a partition of unity argument to carry the result over to M.

As a result, $e^{-t\Delta} : L^2(M) \to H_1(M)$ is continuous. The composition

$$H_0 = L^2 \xrightarrow{e^{-t\Delta}} H_1 \hookrightarrow H_0 = L^2$$

of a continuous operator with a compact operator is compact (check!), and so $e^{-t\Delta} : L^2 \to L^2$ is a compact operator. Moreover, the heat operator is self-adjoint:

$$
\begin{aligned}
\langle e^{-t\Delta}f, g \rangle &= \int_M \left(\int_M e(t, x, y)f(y)dy \right) g(x)dx \\
&= \int_M \left(\int_M e(t, y, x)g(x)dx \right) f(y)dy \\
&= \langle f, e^{-t\Delta}g \rangle, \tag{1.25}
\end{aligned}
$$

by the symmetry of the heat kernel $e(t, x, y)$. By the spectral theorem for self-adjoint compact operators on Hilbert spaces, $L^2(M)$ has an orthonormal basis consisting of eigenfunctions for the heat operator $e^{-t\Delta}$ with eigenvalues $\gamma_i(t) \to 0$ as $i \to \infty$. We write this schematically as

$$e^{-t\Delta} \sim \begin{pmatrix} \gamma_1(t) & & & \\ & \ddots & & \\ & & \gamma_n(t) & \\ & & & \ddots \end{pmatrix}.$$

In order to prove the Hodge theorem, we study the $\gamma_i(t)$. We want to show first of all that they are strictly positive. To do this, we establish the important semigroup property of the heat operator.

Claim: $e^{-t\Delta}e^{-t'\Delta} = e^{-(t+t')\Delta}$.

Since the heat operator is supposed to model heat flow on M from a given initial temperature distribution, the claim is the physically plausible statement that heat flow at time $t+t'$ should be the composition of heat flow up to time t' with heat flow for a further time t.

PROOF: For $t > s > 0$,

$$
\begin{aligned}
e^{-(t-s)\Delta}e^{-s\Delta}f(x) &= e^{-(t-s)\Delta_z}\left(\int_M e(s,z,y)f(y)dy\right)(x) \\
&= \int_M e(t-s,x,z)\left(\int_M e(s,z,y)f(y)dy\right)dz \\
&= \int_M \left(\int_M e(t-s,x,z)e(s,z,y)dz\right)f(y)dy,
\end{aligned}
$$

so $e^{-(t-s)\Delta}e^{-s\Delta}$ has the integral kernel

$$
\int_M e(t-s,x,z)e(s,z,y)dz.
$$

Now $(\partial_t + \Delta_x)(\int_M e(t-s,x,z)e(s,z,y)dz) = 0$, and

$$
\begin{aligned}
\lim_{t\to 0}\int_M \left(\int_M e(t-s,x,z)e(s,z,y)dz\right)f(y)dy \\
= \lim_{t\to 0}\int_M e(t-s,x,z)\left(\lim_{s\to 0}\int_M e(s,z,y)f(y)dy\right)dz \\
= \lim_{t\to 0}\int_M e(t,x,z)f(z)dz \\
= f(x).
\end{aligned}
$$

Here we have used that $s \to 0$ as $t \to 0$ and (1.24) for $e(s,z,y)$. Thus $\int_M e(t-s,x,z)e(s,z,y)$ has the two defining properties of the heat kernel, so by Exercise 25, we must have

$$
e(t,x,y) = \int_M e(t-s,x,z)e(s,z,y)dz.
$$

In other words

$$
e^{-t\Delta} = e^{-(t-s)\Delta}e^{-s\Delta},
$$

which finishes the proof.

Remark: More generally, it is shown in the theory of unbounded operators that a self-adjoint operator D has an associated semigroup $\{e^{-tD}\}$ of self-adjoint operators giving a solution to heat flow [70]. Conversely, it is easy to see that if the semigroup $\{e^{-tD}\}$ has a smooth kernel, this kernel must satisfy (1.24).

We can now show that $\gamma_i(t) > 0$. Since

$$
\begin{aligned}
\langle e^{-t\Delta}f,f\rangle &= \langle e^{-\frac{t}{2}\Delta}e^{-\frac{t}{2}\Delta}f,f\rangle \\
&= \langle e^{-\frac{t}{2}\Delta}f, e^{-\frac{t}{2}\Delta}f\rangle \\
&\geq 0,
\end{aligned}
\tag{1.26}
$$

we certainly have $\gamma_i(t) \geq 0$. If some $\gamma_i(t) = 0$, then there exists $f \neq 0$ with $e^{-t\Delta}f = 0$. As above, we would then have

$$0 = \langle e^{-t\Delta}f, f \rangle = \langle e^{-\frac{t}{2}\Delta}e^{-\frac{t}{2}\Delta}f, f \rangle,$$

which implies $e^{-\frac{t}{2}\Delta}f = 0$. Repeating this argument gives $e^{-\frac{t}{4}\Delta}f = 0$, etc. This yields $f = \lim_{t \to 0} e^{-t\Delta}f = 0$, a contradiction. Thus all the $\gamma_i(t)$ are strictly positive.

We now show that there exist $\lambda_i \in \mathbf{R}$ such that

$$\gamma_i(t) = e^{-\lambda_i t}$$

for all t. There exists an orthonormal basis $\{\omega_i(t)\} \subset L^2(M)$ such that

$$e^{-t\Delta}\omega_i(t) = \gamma_i(t)\omega_i(t).$$

In fact, $\omega_i(t)$ is independent of t, since the equation

$$e^{-s\Delta}e^{-t\Delta} = e^{-(s+t)\Delta} = e^{-(t+s)\Delta} = e^{-t\Delta}e^{-s\Delta}$$

implies that the operators $e^{-t\Delta}$ can be simultaneously diagonalized for all t. (Check that this result from linear algebra applies in the current infinite dimensional case.) Thus $e^{-t\Delta}\omega_i = \gamma_i(t)\omega_i$, and

$$
\begin{aligned}
0 &= (\partial_t + \Delta)(e^{-t\Delta}\omega_i) \\
&= (\partial_t + \Delta)(\gamma_i(t)\omega_i) \\
&= \dot{\gamma}_i\omega_i + \gamma_i\Delta\omega_i,
\end{aligned}
$$

where $\dot{\gamma}_i$ denotes $\partial_t \gamma_i$. Since we know $\gamma_i > 0$, we may write

$$\Delta\omega_i = -\frac{\dot{\gamma}_i}{\gamma_i}\omega_i. \tag{1.27}$$

The left hand side of this equation is independent of t, which implies

$$-\frac{\dot{\gamma}_i}{\gamma_i} = C,$$

with C a constant which depends only on i and not on t. Thus $\gamma_i(t) = C' \cdot e^{-\lambda_i t}$, for some real number λ_i. As $t \to 0$, $e^{-t\Delta}$ goes to the identity. This forces $\gamma_i(t) \to 1$ as $t \to 0$, and so we must have $C' = 1$ and $\lambda_i \geq 0$. In conclusion,

$$\gamma_i(t) = e^{-\lambda_i t}, \tag{1.28}$$

and

$$e^{-t\Delta} \sim \begin{pmatrix} e^{-\lambda_1 t} & & \\ & e^{-\lambda_2 t} & \\ & & \ddots \end{pmatrix}$$

with respect to the basis $\{\omega_i\}$ in $L^2(M, g)$. By (1.27) and (1.28), we have

$$\Delta\omega_i = -\frac{-\lambda_i e^{-\lambda_i t}}{e^{-\lambda_i t}}\omega_i = \lambda_i\omega_i.$$

Thus, with respect to the same basis $\{\omega_i\}$, we have the desired diagonalization of the Laplacian:

$$\Delta \sim \begin{pmatrix} \lambda_1 & & \\ & \lambda_2 & \\ & & \ddots \end{pmatrix}.$$

Remarks: (1) We can also show that the eigenvalues λ_i of Δ are nonnegative by the equation

$$\langle \Delta f, f \rangle = \langle \delta df, f \rangle = \langle df, df \rangle \geq 0.$$

The 0-eigenspace of Δ is easy to determine. If $\Delta f = 0$, then

$$0 = \langle \Delta f, f \rangle = \langle df, df \rangle.$$

Therefore, $df = 0$, and f is a constant function. (This is the only place in this subsection where we use that M is connected.) Thus 0 is an eigenvalue of multiplicity one. Since $e^{-t\Delta}$ is a compact operator, its eigenvalues $e^{-t\lambda_i}$ all have finite multiplicity and accumulate only at 0. Thus we may list the eigenvalues of Δ (repeating eigenvalues according to their multiplicity):

$$0 = \lambda_1 < \lambda_2 \leq \lambda_3 \leq \lambda_4 \leq \ldots \uparrow \infty.$$

Since the eigenvalues go to infinity, Δ is always an unbounded operator on $L^2(M)$.

(2) The eigenfunctions ω_i are the generalization of the trigonometric polynomials on S^1. (More precisely, the ω_i generalize the real eigenfunctions $\{\cos(n\theta), \sin(n\theta): n = 0, 1, 2, \ldots\}$ of the Laplacian on S^1.) In fact, for $f \in L^2(M, g)$, we may write

$$f = \sum_i a_i\omega_i; \quad a_i = \langle f, \omega_i \rangle = \int_M f \cdot \omega_i \text{ dvol}.$$

Since $e^{-t\Delta}$ is bounded on L^2, we have

$$\begin{aligned} e^{-t\Delta}f &= e^{-t\Delta}\sum a_i\omega_i \\ &= \sum a_i e^{-t\Delta}\omega_i \\ &= \sum a_i e^{-\lambda_i t}\omega_i. \end{aligned}$$

In particular, as $t \to \infty$, $e^{-t\Delta}f \to a_1\omega_1$. Now ω_1 is a constant function suitably normalized,

$$\omega_1 = 1/\|1\| = 1/(\int_M 1 \cdot 1 \text{ dvol})^{\frac{1}{2}} = \frac{1}{\sqrt{\text{vol}(M)}},$$

and so

$$
\begin{aligned}
a_1\omega_1 &= \langle f, \omega_1 \rangle \omega_1 \\
&= \left(\int_M f \cdot \frac{1}{\sqrt{\mathrm{vol}(M)}} \, \mathrm{dvol} \right) \frac{1}{\sqrt{\mathrm{vol}(M)}} \\
&= \frac{\int_M f \, \mathrm{dvol}}{\mathrm{vol}(M)}.
\end{aligned}
$$

Thus, just as on S^1, heat flow on M sends a function to its average value as $t \to \infty$.

(3) Consider the formal sum $\sum_i e^{-\lambda_i t} \omega_i(x) \omega_i(y)$. Formally,

$$
\begin{aligned}
\int_M \left(\sum_i e^{-\lambda_i t} \omega_i(x) \omega_i(y) \right) f(y) dy &= \sum_i e^{-\lambda_i t} \omega_i(x) \int_M \omega_i(y) f(y) dy \\
&= \sum_i e^{-\lambda_i t} \omega_i(x) a_i.
\end{aligned}
$$

This shows that formally the heat kernel has the expression $\sum e^{-\lambda_i t} \omega_i(x) \omega_i(y)$. We'll show later that this sum does indeed converge nicely to $e(t, x, y)$.

To sum up the results so far, we state the Hodge theorem for functions:

Theorem 1.29 (Hodge Theorem for Functions) *Let (M, g) be a compact connected oriented Riemannian manifold. There exists an orthonormal basis of $L^2(M, g)$ consisting of eigenfunctions of the Laplacian. All the eigenvalues are positive, except that zero is an eigenvalue with multiplicity one. Each eigenvalue has finite multiplicity, and the eigenvalues accumulate only at infinity.*

Exercise 28: *Let $\mathbf{Z} \oplus \mathbf{Z}$ denote the subgroup of the (additive) group $\mathbf{R}^2$ generated by $(1, 0)$ and $(0, 1)$.*

(i) Show that each element of $\mathbf{Z} \oplus \mathbf{Z}$ acts isometrically on $\mathbf{R}^2$; i.e. for $g \in \mathbf{Z} \oplus \mathbf{Z}$, the translation map $T_g : \mathbf{R}^2 \to \mathbf{R}^2$ given by $T_g(v) = v + g$ preserves the standard inner product on (each tangent space to) $\mathbf{R}^2$. As in Exercise 5, this means that $(T_g)_* : T_v \mathbf{R}^2 \to T_{v+g} \mathbf{R}^2$ is an isometry. Conclude that the quotient manifold $T^2 = \mathbf{R}^2 / (\mathbf{Z} \oplus \mathbf{Z})$ (the two-torus) has a well defined metric such that the projection map from $\mathbf{R}^2$ to $\mathbf{R}^2 / (\mathbf{Z} \oplus \mathbf{Z})$ is a local isometry.

(ii) Let θ, ψ denote the usual angular coordinates on the torus T^2. Compute the Laplacian on T^2 in these coordinates. Compute the eigenvalues and the eigenfunctions of the Laplacian.

Exercise 29: *Compute the eigenfunctions and the eigenvalues for the Laplacian on S^2 with the standard metric. Hint: This exercise is harder than Exercise 28. You may want to read [4] or [24].*

Remark: The Hodge theorem is similar to the basic theorem of Sturm-Liouville theory: the existence of an orthonormal basis of eigenfunctions for certain second order differential operators on an interval with Dirichlet or Neumann boundary conditions. In fact, the Hodge theorem as stated above is valid for manifolds with boundary, provided one imposes Dirichlet or Neumann boundary conditions. The proof is as before, once one checks that the results of §1.3.1 carry over to functions on $\overline{\Omega}$ with these boundary conditions (which is straightforward) and that a heat kernel exists (which is rather delicate). Moreover, the Hodge theorem is valid for a class of operators, the so-called elliptic operators, which include the Laplacians we deal with. Elliptic operators on the interval are precisely the operators treated in Sturm-Liouville theory, so Hodge theory does extend these classical results.

1.3.3 The Hodge Theorem for Differential Forms

In this section we define the Laplacian on k-forms and prove the corresponding version of the Hodge theorem.

In §1.2.3, we computed the adjoint δ^{k+1} for exterior differentiation d^k on k-forms, so it might seem reasonable to define the Laplacian on k-forms to be $\delta^{k+1}d^k$. This doesn't work, for various reasons. First of all, this definition makes no sense if $k = n$. Secondly, the kernel of $\delta^{k+1}d^k$ is infinite dimensional, as it contains the image of d^{k-1}, and so any heat operator associated to this operator could not be compact on $L^2\Lambda^k$. All this shows that our first guess for the Laplacian is not correct, but it does not tell us what the correct definition should be.

We'll now try to motivate our definition; the reader should not take this attempt too seriously, as the real motivation comes from the discussion in §1.5.

Exercise 30: *Let* dim $M = n$. *Show that* $*\Delta = d^{n-1}\delta^n *$.

From this exercise, we see that $d^{n-1}\delta^n$ is isospectral to Δ, since the star operator takes eigenfunctions of one operator to eigenfunctions of the other. Thus the conclusions of the Hodge theorem for functions are valid for the operator $d^{n-1}\delta^n$ on n-forms. This indicates that $d^{n-1}\delta^n$ might as well be the Laplacian on n-forms, and that a term similar to it should occur in the Laplacian on k-forms.

Definition: *The* Laplacian on k-forms *on a Riemannian manifold is given by*

$$\Delta^k = \delta^{k+1}d^k + d^{k-1}\delta^k.$$

To prove the Hodge theorem for these Laplacians, we need to rework the analytic preliminaries of §1.3.1. Defining Sobolev spaces of k-forms is not difficult. Given a cover $\{U_i, \varphi_i\}$ of (M, g) with subordinate partition of unity $\{\rho_i\}$

as before, we define the Sobolev s-norm of a compactly supported k-form u to be the s-norm of the function $g(u, u)^{1/2}$ on M:

$$\|u\|_s = (\sum_i \|\rho_i g(u,u)^{1/2} \circ \varphi_i\|_s^2)^{1/2}.$$

The completion of $C_c^\infty \Lambda^k T^* M$ with respect to this norm is denoted $H_s \Lambda^k M$.

Exercise 31: *Show that if M is compact, then $H_s \Lambda^k M$ is independent of the choices of cover of M, partition of unity, and Riemannian metric on M. What happens if M is noncompact?*

The argument that the embedding theorem and the compactness theorem hold for Sobolev spaces of forms on compact manifolds is exactly the same as for functions; see Exercise 24. Similarly, the work in Chapter 3 showing the existence of a heat kernel can be generalized to produce a heat kernel and heat operator for the Laplacian on forms; this is discussed in §3.2.

Here we should state precisely what we mean by a heat kernel for forms. A so-called *double form* $e(t, x, y) \in \mathbf{R}^+ \times \Lambda^k T_x^* M \otimes \Lambda^k T_y^* M$ which is smooth in t, x, y is called the heat kernel for the Laplacian on k-forms if

(1) $(\partial_t + \Delta_x^k)e(t, x, y) = 0$,

(2) $\lim_{t \to 0} \int_M \langle e(t, x, y), \omega(y) \rangle_y \, dy = \omega(x)$, for all smooth k-forms ω.

Note that the smoothness of $e(t, x, y)$ is equivalent to saying that $e(t, x, y)$ is a smooth section of $\mathbf{R} \times (\Lambda^k T^* M \otimes \Lambda^k T^* M)$ as a bundle over $M \times M$, where we consider $\mathbf{R}$ as the trivial bundle over $M \times M$. In (2), the pointwise inner product and integration are in the y variable, leaving a form evaluated at x; equivalently, the integral may be written $\int_M e(t, x, y) \wedge *_y \omega(y)$, where $*_y$ indicates that we take the Hodge star and integrate with respect to y. To be concrete, let $I = (i_1, \ldots, i_k)$ be a multi-index and let $dx^I = dx^{i_1} \wedge \ldots \wedge dx^{i_k}$. If we write $e(t, x, y) = f_{IJ}(t, x, y)dx^I \otimes dy^J$ in a coordinate patch U, then for all $\omega = \omega_I dx^I$ with support in U, we have

$$\int_M \langle e(t, x, y), \omega(y) \rangle_y \, dy = \left(\int_M \langle f_{IJ}(t, x, y)dy^J, \omega_K(y)dy^K \rangle \, \mathrm{dvol}(y) \right) dx^I,$$

which is indeed in $\Lambda^k T_x^* M$. Given these analytic facts, the proof of the Hodge theorem for functions carries directly over for the Laplacian on forms. The one crucial difference is that the kernel of Δ^k need not be one dimensional. For completeness, we summarize this discussion:

Theorem 1.30 (Hodge Theorem for Forms) *Let (M, g) be a compact connected oriented Riemannian manifold. There exists an orthonormal basis of $L^2 \Lambda^k(M, g)$ consisting of eigenforms of the Laplacian on k-forms Δ^k. All the eigenvalues are nonnegative. Each eigenvalue has finite multiplicity, and the eigenvalues accumulate only at infinity.*

In the next sections, we will discuss the topological significance of the kernel of Δ^k.

Exercise 32: *Compute a basis of $L^2\Lambda^1(T^2)$ consisting of eigenforms for Δ^1 with respect to the standard metric of Exercise 21. Compute the corresponding eigenvalues.*

It should seem that we are not working very hard in the last exercise, and the reader may wish to see eigenfunctions and eigenforms for the Laplacian for more interesting Riemannian manifolds. However, if one tries to compute the eigenfunctions for the Laplacian on T^2 with the metric induced from the standard embedding of T^2 in $\mathbf{R}^3$, one soon confronts a nasty looking linear PDE whose solutions are not obvious. Trying to compute the eigenforms/eigenvalues for the Laplacian on one-forms on this manifold is even harder, as one encounters a system of linear PDEs. Thus it is usually impossible to compute the eigenforms/eigenvalues explicitly except for very special Riemannian manifolds, and the study of the spectrum of the Laplacians concentrates instead on determining what information can be read off from the spectrum without explicit knowledge of the spectrum. As we will see, it is remarkable how much geometric and topological information can be decoded from the spectrum.

1.3.4 Regularity Results

So far we have derived the L^2 theory for the Laplacians on forms. Although this is the easiest theory from the analytic point of view, from the point of view of differential topology we want to know as much as possible about the C^∞ theory. This requires some adjustments. First of all, although we have been vague about the domain of the Laplacian on forms on a compact manifold, it is certainly plausible (and in fact true) that its domain contains all C^2 forms. The heat equation methods used in the proof of the Hodge theorem allow us to show that the eigenforms of the Laplacian $\Delta = \Delta^k$ on k-forms are smooth, for all k.

Theorem 1.31 *Let $\omega \in C^2\Lambda^k T^* M$ satisfy $\Delta\omega = \lambda\omega$. Then $\omega \in C^\infty\Lambda^k T^* M$.*

Remark: It should not be too surprising that this regularity result also holds for noncompact manifolds, since the degree of differentiability of a function is a local property. We omit the proof here to avoid some technicalities. In particular, this regularity theorem generalizes a classical theorem of Weyl which states that harmonic functions on Euclidean space are smooth.

PROOF. If $\Delta\omega = \lambda\omega$, then

$$e^{-t\Delta}\omega = e^{-t\lambda}\omega.$$

Since we will show that the heat operator on forms has a smooth kernel, the argument at the beginning of §1.3.2 shows that $e^{-t\Delta}\omega$ is smooth. The last equation implies that ω is itself smooth.

Using this regularity theorem, we can pass from the decomposition of the space of L^2 k-forms to a more useful decomposition of the space of smooth forms. First of all, if ω is in the kernel of $\Delta = \Delta^k$, then

$$0 = \langle \Delta\omega, \omega \rangle = \langle d\delta\omega, \omega \rangle + \langle \delta d\omega, \omega \rangle = \langle \delta\omega, \delta\omega \rangle + \langle d\omega, d\omega \rangle,$$

and so $d\omega = \delta\omega = 0$. Conversely, it is immediate that $d\omega = \delta\omega = 0$ implies $\Delta\omega = 0$.

Now consider $d : C^\infty\Lambda^{k-1}T^*M \rightarrow C^\infty\Lambda^k T^*M$ and $\delta : C^\infty\Lambda^{k+1}T^*M \rightarrow C^\infty\Lambda^k T^*M$ as operators on smooth forms. We have

$$C^\infty\Lambda^k T^*M \supset \operatorname{Ker} \Delta \oplus \operatorname{Im} d \oplus \operatorname{Im} \delta, \qquad (1.32)$$

where $\oplus$ indicates orthogonality with respect to the Hodge inner product. For example, if $\omega \in \operatorname{Ker} \Delta$, we have $\langle \omega, d\alpha \rangle = \langle \delta\omega, \alpha \rangle = 0$; the other orthogonality relations follow similarly. Our goal in the rest of this section is to show that the two sides of (1.32) are in fact equal.

The form of (1.32) suggests that we work with the operator $D = d + \delta$, thought of as an operator on $C^\infty\Lambda^*$, the space of smooth forms of mixed degree. The operator D has the advantage of being a first order differential operator and is a "square root" of the Laplacian on $C^\infty\Lambda^*$: $D^2 = \Delta$. More importantly, our analysis of D will lead to the Chern-Gauss-Bonnet theorem in Chapter 4.

Note that since $\operatorname{Im} d \perp \operatorname{Im} \delta$ and $\operatorname{Ker} \Delta = \operatorname{Ker} d \cap \operatorname{Ker} \delta$, it suffices to show that

$$C^\infty\Lambda^* = \operatorname{Ker} D \oplus \operatorname{Im} D. \qquad (1.33)$$

We now just write H_s, L^2, C^∞ for $H_s\Lambda^*, L^2\Lambda^*, C^\infty\Lambda^*$, the spaces of forms of mixed degree with the Sobolev norm, L^2 norm and smoothness criterion, respectively.

The first step is Gårding's inequality, also called the basic elliptic estimate:

Theorem 1.34 *For each $s \in \mathbf{Z}^+ \cup \{0\}$, there exists a positive constant $C = C_s$ such that*

$$\|\omega\|_s \leq C(\|\omega\|_{s-1} + \|D\omega\|_{s-1}),$$

for all $\omega \in H_s$.

D is a bounded operator from a dense subset of H_s to H_{s-1}, and so extends to a bounded operator on all of H_s. Thus the right hand side of Gårding's inequality makes sense. Notice that the reverse inequality, with a different C, is immediate from the definition of H_s. Thus the two sides of the inequality define equivalent norms for H_s.

A clever proof of Theorem 1.34 is given in Theorem 2.46. Standard proofs for general elliptic operators are in [24], [30]. For ω a function, the crucial property of D in the standard proof is that the coefficients (g^{ij}) which occur in the second order term of the Laplacian D^2 form a positive definite matrix. Since the top order term of the Laplacian on forms is the top order term of the

Laplacian on functions times an identity matrix (see Exercise 33), the argument for functions extends to the case of forms.

Exercise 33: *Let $\omega = a_I dx^I$ be a k-form on a Riemannian manifold M. Show that $\Delta^k \omega = (\Delta^0 a_I) dx^I + $ (lower order terms). Hint: Write $\omega = b_I \theta^I$, where $\{\theta^i(x)\}$ is an orthonormal basis of $T_x^* M$ in a neighborhood of a fixed point x_0. It suffices to prove the exercise in the θ "coordinates." If $\{X_i\}$ is the basis of $T_x M$ dual to $\{\theta^i\}$, note that for any function f we have $df = X_i(f)\theta^i$ and the second order term in Δf is $-\sum_i X_i(X_i(f))$. Now compute the second order term of $\Delta^k \omega$ by differentiating b_I as much as possible. Warning: this is a long computation, so proceed only if you want a deeper appreciation of Theorem 2.46.*

In fact, as the next series of Exercises show, one can derive the Hodge theorem from the Sobolev embedding theorem, the compactness theorem, and Gårding's inequality, without any use of the heat equation. This is the usual elliptic proof of the Hodge theorem.

Exercise 34: *(i) Show that D is injective as a map $D : H_s^\perp \equiv H_s \cap (\text{Ker } D)^\perp \to H_{s-1}$, where $\perp$ indicates orthogonal complement with respect to the Hodge metric on L^2 forms. Thus we can define $D^{-1} : R(D) \to H_s^\perp$, for $R(D)$ the range of D on $H_s^\perp$.*

(ii) Show that $D^{-1} : L^2 \cap (\text{Ker } D)^\perp \to H_1 \hookrightarrow L^2$ is bounded. Hint: Since the spectrum of D^{-1} consists of the reciprocals of the nonzero spectrum of D, it suffices to show that the spectrum of D restricted to $(\text{Ker } D)^\perp$ is bounded away from zero. If there exist ω_i with $\|\omega_i\|_0 = 1$ and $\|D\omega_i\|_0 \to 0$, show that $d\omega_i \to 0, \delta\omega_i \to 0$ in L^2. This implies that ω_i forms a bounded sequence in H_1 by Gårding, and so there exists ω with $\omega_i \to \omega$ in L^2. Thus $\omega \in (\text{Ker } D)^\perp$. But $\langle D\omega_i, \theta \rangle = \langle \omega_i, D\theta \rangle \to 0$ for all $\theta \in L^2$ implies that $\omega \in (\text{Im } D)^\perp = \text{Ker } D$. Thus $\omega \in (\text{Ker } D) \cap (\text{Ker } D)^\perp = \{0\}$, which contradicts $\|\omega\|_0 = 1$.

(iii) Use Gårding's inequality and induction to show that $D^{-1} : R(D)(\subset H_{s-1}) \to H_s^\perp$ is bounded for all s.

(iv) In particular, by the compactness theorem, $D^{-1} : L^2 \cap (\text{Ker } D)^\perp \to L^2$ is a compact operator. Now apply the spectral theorem for self-adjoint operators to D^{-1} to give another proof of the Hodge theorem.

We can also use Gårding's inequality to extend our regularity results.

Corollary 1.35 *If $D\alpha = \beta$ for k-forms α, β and if $\alpha, \beta \in H_s$, then in fact $\alpha \in H_{s+1}$. In particular, if $\beta \in C^\infty$ and $\alpha \in C^1$, then $\alpha \in C^\infty$.*

PROOF. This follows from $\|\alpha\|_{s+1} \leq C(\|\alpha\|_s + \|\beta\|_s)$.

We will now restate this last result. Recall that the closure $\overline{D}$ of D is defined by taking the closure of the graph of D in L^2: i.e. $\overline{D}\alpha = \beta$ if there exist $\alpha_i \in L^2$ with $\alpha_i \to \alpha$ in L^2 and $D\alpha_i \to \beta$ in L^2. Note that since D is unbounded, we cannot write $D\alpha = \beta$.

Corollary 1.36 *If $\overline{D}\alpha = \beta$ with $\beta \in C^\infty$, then $\alpha \in C^\infty$ and $D\alpha = \beta$.*

PROOF. Take $\alpha_i \to \alpha$ with $D\alpha_i \to \beta$. By the definition of the domain of D (which we have not made precise but which certainly contains the space of C^1 forms), we may assume that $\alpha_i, \alpha \in H_1^\perp$. D^{-1} is bounded on L^2 by the last exercise, so $\alpha_i \to D^{-1}\beta$, and hence $\alpha = D^{-1}\beta$ in L^2. But $D^{-1}\beta \in H_1$ by the last exercise, and so $\alpha \in H_1$. Now D is defined on all of H_1, and so $\overline{D}\alpha = D\alpha$. The last corollary now implies that $\alpha \in C^\infty$.

We now wish to determine $R(\overline{D}) \subset L^2$. You should check that $R(\overline{D}) \perp$ Ker D. We claim that in fact $R(\overline{D}) = (\text{Ker } D)^\perp$. Since D^{-1} is bounded on $(\text{Ker }\Delta)^\perp$ by the last exercise, it is easy to show that $R(\overline{D})$ is closed. If $R(\overline{D})$ is not all of $(\text{Ker } D)^\perp$, then there exists $0 \neq \alpha \in (\text{Ker } D)^\perp$ with $\alpha \perp R(D)$ (check this!). In other words, $\langle \alpha, D\theta \rangle = 0$ for all $\theta \in C^\infty$. We can find $\alpha_i \in C^\infty$ with $\alpha_i \to \alpha$ in L^2. Then $\langle D\alpha_i, \theta \rangle \to 0$ in L^2 for all $\theta \in C^\infty$ implies that $D\alpha_i \to 0$ in L^2. By definition this means $\overline{D}\alpha = 0$, so by the last corollary $D\alpha = 0$, which is a contradiction.

Thus we have shown that

$$L^2\Lambda^* = \text{Ker } D \oplus R(\overline{D}).$$

Now take $\omega \in C^\infty$, and let $P\omega$ denote its orthogonal projection into Ker D, which consists of smooth forms by the last corollary. Then $\omega - P\omega$ is smooth and in the range of $\overline{D}$. By the last corollary, we can write $\omega - P\omega = D\beta$ for some $\beta \in C^\infty$. This proves (1.33), which we state as follows:

Theorem 1.37 (Hodge Decomposition Theorem)

$$C^\infty\Lambda^k T^* M = \text{Ker } \Delta^k \oplus \text{Im } d^{k-1} \oplus \text{Im } \delta^{k+1}.$$

Finally, we show that the decomposition of L^2 into eigenspaces of D extends to a decomposition of H_s; this technical result will be used later.

Proposition 1.38 *Let $\{\omega_i\}$ be an orthonormal basis of L^2 with $D\omega_i = \lambda_i\omega_i$. If $\omega \in H_s$, for $s \in \mathbf{Z}^+$, has $\omega = \sum_i a_i\omega_i$ with equality in L^2, then $\omega = \sum_i a_i\omega_i$ with equality in H_s.*

PROOF. Since $\omega \in H_s$, we have $\omega \in H_1$, and so $D\omega = \sum_i b_i\omega_i$ in L^2, where

$$b_i = \langle D\omega, \omega_i \rangle = \langle \omega, D\omega_i \rangle = \lambda_i\langle \omega, \omega_i \rangle = \lambda_i a_i.$$

By Exercise 34(iii), $D^{-1} : L^2 \cap (\text{Ker } D)^\perp \to H_1$ is bounded, and clearly $D^{-1}\omega_i = \lambda_i^{-1}\omega_i$ for $\lambda_i \neq 0$. Thus $\omega - P\omega = D^{-1}D\omega = \sum_i' a_i\omega_i$ in H_1. Here $P\omega$ is the L^2 projection of ω into Ker D, and $\sum_i'$ indicates that we sum over those i with $\lambda_i \neq 0$. Since $P\omega$ is a finite sum $\sum_i'' a_i\omega_i$ over i with $\lambda_i = 0$, we get that $\omega = \sum_i a_i\omega_i$ in H_1.

Assume by induction that for some fixed $s > 1$, $\omega \in H_s$ and that we have shown $\omega = \sum_i a_i \omega_i$ with equality in H_{s-1}. Moreover, assume by induction that the proposition is true for all $\omega \in H_k$, $k < s$. For any N, Gårding yields

$$\left\| \omega - \sum_{i=1}^{N} a_i \omega_i \right\|_s \leq C \left(\left\| \omega - \sum_{i=1}^{N} a_i \omega_i \right\|_{s-1} + \left\| D(\omega - \sum_{i=1}^{N} a_i \omega_i) \right\|_{s-1} \right). \quad (1.39)$$

Since we know $D\omega = \sum_i \lambda_i a_i \omega_i$ in L^2 and $D\omega \in H_{s-1}$, by the induction hypothesis we get the same equality in H_{s-1}. Thus the right hand side of (1.39) goes to zero as $N \to \infty$. Thus $\omega = \sum_i a_i \omega_i$ in H_s.

The results of this subsection can easily be extended to the Laplacian D^2 to show that $\Delta \alpha = \beta$ has a solution iff $\beta \in (\text{Ker } \Delta)^{\perp}$ and that the solution always has degree of smoothness two more than β. These results should be viewed as generalizations of classical results in potential theory regarding the solvability of Laplace's equation $\Delta \alpha = \beta$ for domains in Euclidean space (although a more exact analogue would be solving Laplace's equation on manifolds with boundary). In turn, all our results can be generalized to a wider class of operators, the elliptic operators, on compact manifolds via the calculus of pseudo-differential operators. For this topic we recommend [30].

1.4 De Rham Cohomology

In this section we will see how to use differential forms to detect topological properties of smooth manifolds. Our discussion of de Rham cohomology is based on [32], [64, Vol. I]; a much more thorough treatment is in [11].

As a basic example, we consider the two-torus $T^2 = S^1 \times S^1$. We parametrize points on the torus via the usual angular coordinates (θ, ψ), $\theta, \psi \in [0, 2\pi)$, of the two circles. Note that strictly speaking these coordinates give a coordinate chart only on a dense subset of T^2.

Loosely speaking, among all oriented closed surfaces, the torus is characterized by having one hole. In other words, among all such surfaces only the torus has exactly two "really different" noncontractible loops, $a : [0, 2\pi] \to T^2$, $b : [0, 2\pi] \to T^2$, given by $a(t) = (t, 0), b(t) = (0, t)$. Here "really different" means that a and b cannot be deformed into each other, and that any other loop on T^2 can be deformed into a loop which travels n times around a and m times around b for some $m, n \in \mathbf{Z}$. Using the map $(n, m) \mapsto na + mb$, we see that the group $\mathbf{Z} \oplus \mathbf{Z}$ appears naturally as an algebraic description of the one hole in the torus. The reader probably knows that this group is called *the fundamental group* of the torus.

Exercise 35: *Compute the fundamental group of the g-holed torus. Hint: the group has 2g generators and one relation.*

Now we want to construct differential forms which detect the difference between the loops a, b, but which are insensitive to smooth deformations of these loops. As candidates, we take the one-forms $d\theta,\; d\psi \in \Lambda^1 T^2$, which satisfy

$$\int_a d\theta = \int_b d\psi = 2\pi, \quad \int_b d\theta = \int_a d\psi = 0.$$

Moreover, if $F : [0,1] \times [0,1] \to T^2$, $F(0,t) = a(t)$, is a smooth deformation of a to the loop $a' : t \mapsto F(1,t)$, then by Stokes' theorem

$$\int_a d\theta = \int_{a'} d\theta + \int_{\mathrm{Im}\,F} d(d\theta) = \int_{a'} d\theta.$$

(Note that $\int_{\mathrm{Im}\,F}$ is actually an integral over $[0,1] \times [0,1]$ and so is well defined even if the image of F is messy.) Thus $d\theta, d\psi$ satisfy the conditions we specified.

Exercise 36: *Why is the following argument invalid? Since the image of the loop a is a one-manifold without boundary, by Stokes' theorem*

$$\int_a d\theta = \int_{\partial a = \emptyset} \theta = 0.$$

The key property of $d\theta$ used in the preceding computations was that $d\theta$ is *closed*, i.e. $d(d\theta) = 0$. However, for any $f \in C^\infty(T^2)$, $d\theta + df$ is also closed, with

$$\int_a (d\theta + df) = \int_a d\theta + \int_{\partial a} f = \int_a d\theta,$$

and similarly for the integral over b. Thus as far as integration over loops is concerned, the forms $d\theta$ and $d\theta + df$ are indistinguishable; similar remarks hold for $d\psi$. Since the *exact* one-forms $\{df : f \in C^\infty(T^2)\}$ are included in the closed one-forms, we can quotient out the ambiguity by setting the *first de Rham cohomology group* of the torus to be

$$H^1_{dR}(T^2) = \{\omega \in \Lambda^1 T^2 : d\omega = 0\} / \{df : f \in C^\infty(T^2)\}.$$

Of course, $H^1_{dR}(T^2)$ is naturally a real vector space and so a group only in a trivial sense, but the terminology will become clearer soon.

The basic information contained in this vector space is its dimension.

Claim: dim $H^1_{dR}(T^2) = 2$.

PROOF. Let $\omega = \alpha d\theta + \beta d\psi$ be a closed one-form. A direct calculation gives

$$\frac{\partial \alpha}{\partial \psi} = \frac{\partial \beta}{\partial \theta}. \tag{1.40}$$

This equation pulls back to $\mathbf{R}^2$ as in Exercise 28, where we now consider $\alpha, \beta, \theta, \psi$ as periodic functions on $\mathbf{R}^2$. Then (1.40) is just the *integrability condition*, i.e. the necessary and sufficient condition, to solve the equations

$$\frac{\partial f}{\partial \theta} = \alpha, \quad \frac{\partial f}{\partial \psi} = \beta$$

for $f \in C^\infty(\mathbf{R}^2)$. This is equivalent to solving

$$df = \omega \tag{1.41}$$

in $\mathbf{R}^2$. If (1.41) were valid on T^2, we would have $H^1_{dR}(T^2) = 0$. However, (1.41) is valid on T^2 iff f defines a function on the quotient $T^2 = \mathbf{R}^2/(\mathbf{R} \oplus \mathbf{R})$. In other words, f must be periodic: $f(x+2\pi n, y+2\pi m) = f(x,y)$ for all $n, m \in \mathbf{Z}$.

If α, β are constants not both zero, then $f(x,y) = \alpha x + \beta y + C$ for some constant C. Hence f does not descend to a function on T^2. This shows that in this case $\omega \neq df$ on T^2, and thus that the dimension of $H^1_{dR}(T^2)$ is at least two – specifically, the equivalence classes $[d\theta], [d\psi]$ are nonzero.

Now we want to study what happens when α, β have no constant part. To make sense of this, we write the Fourier expansions of α, β:

$$\alpha(\theta, \psi) = \sum_{n,m \in \mathbf{Z}} a_{n,m} e^{in\theta} e^{im\psi}, \quad \beta(\theta, \psi) = \sum_{n,m \in \mathbf{Z}} b_{n,m} e^{in\theta} e^{im\psi},$$

and assume that $a_{0,0} = b_{0,0} = 0$. (This Fourier expansion is the solution to Exercise 28(ii).)

Exercise 37: *For the standard metric on T^2, show that this assumption on α, β is equivalent to assuming that α, β are orthogonal to the constants in $L^2(T^2)$.*

Under this assumption, one directly calculates that

$$\int_a \alpha = \int_b \alpha = \int_a \beta = \int_b \beta = 0. \tag{1.42}$$

Lifting back to $\mathbf{R}^2$ and using the explicit solution

$$f(x_0, y_0) = f(0,0) + \int_{(0,0)}^{(x_0,0)} \alpha(x,0)\, dx + \int_{(x_0,0)}^{(x_0,y_0)} \beta(x_0, y)\, dy,$$

we get

$$f(x_0 + 2\pi n, y_0 + 2\pi m) - f(x_0, y_0)$$
$$= \int_{(x_0,0)}^{(x_0+2\pi n,0)} \alpha(x,0)\, dx - \int_{(x_0,0)}^{(x_0,y_0)} \beta(x_0, y)\, dy$$
$$+ \int_{(x_0+2\pi n,0)}^{(x_0+2\pi n, y_0+2\pi m)} \beta(x_0 + 2\pi n, y)\, dy$$
$$= m \int_{(x_0,y_0)}^{(x_0,y_0+2\pi)} \beta(x_0, y)\, dy$$
$$+ n \int_{(x_0,0)}^{(x_0+2\pi,0)} \alpha(x,0)\, dx,$$

by the periodicity of $\alpha(x,y), \beta(x,y)$. Now (1.42) shows the right hand side of the last equation vanishes, since the integrals in (1.42) are unchanged under deformation of a, b to vertical and horizontal circles, respectively. This shows that f is periodic and hence $df = \omega$ on T^2. As a result, $H^1_{dR}(T^2)$ is precisely two dimensional with basis $[d\theta], [d\psi]$.

Exercise 38: *Modify this proof to compute $H^1_{dR}(S^1)$.*

Thus the fact that the fundamental group of the torus is a free **Z**-module with two generators is mirrored by our computation that the first de Rham cohomology group is two dimensional. However, the reader should be aware that there is a loss of information in passing from the fundamental group $\pi_1(M)$ of a manifold M to de Rham cohomology $H^1_{dR}(M)$, the quotient of the closed one-forms on M by the exact one-forms. For example, the fundamental group of two dimensional real projective space is $\mathbf{Z}_2$, whereas the first de Rham cohomology group vanishes. In fact, if $[\pi_1(M), \pi_1(M)]$ denotes the commutator subgroup of $\pi_1(M)$, it turns out that $H^1(M) \cong (\pi_1(M)/[\pi_1(M), \pi_1(M)]) \otimes \mathbf{R}$. Any torsion element in $\pi_1/[\pi_1, \pi_1]$ becomes trivial after tensoring with $\mathbf{R}$, so in passing from π_1 to H^1_{dR} we lose the ability to detect both the non-abelian character of π_1 and the existence of elements of finite order in π_1. On the other hand, it is much easier to compute the de Rham cohomology of a manifold than to compute the fundamental group and its higher dimensional analogues, the homotopy groups.

Just as for the loops a, b on T^2, the space of closed k-forms modulo exact k-forms detects deformation classes of k-dimensional submanifolds, and so should contain topological information.

Definition: *The kth de Rham cohomology group of a manifold M, $H^k_{dR}(M)$, is given by:*

$$H^k_{dR}(M) = \{\omega \in \Lambda^k : d\omega = 0\}/\{d\theta : \theta \in \Lambda^{k-1}\}.$$

The dimension β^k of $H^k_{dR}(M)$ is called the kth Betti number.

Since the spaces of closed and exact forms are both infinite dimensional, it is a nontrivial fact that $\beta^k < \infty$ for all k whenever M is compact; cf. §1.5.

For example, $H^0_{dR}(M) = \{f \in \Lambda^0 : df = 0\}$, which means that f must be constant on each connected component of M. Thus if M has q components, we have $H^0_{dR}(M) \cong \mathbf{R}^q$ and $\beta^0 = q$. In other words, the zeroth Betti number has a topological significance, albeit not a very exciting one.

To explain the topological significance of the higher Betti numbers, we will show that if M is smoothly homotopy equivalent to N, then $H^k_{dR}(M) \cong H^k_{dR}(N)$. Recall that homotopy equivalence means that there exist maps

$$f : M \to N, \ g : N \to M, \ F : [0,1] \times M \to M, \ G : [0,1] \times N \to N,$$

with

$$F(0, x) = (gf)(x), \ F(1, x) = x, \ G(0, y) = (fg)(y), \ G(1, y) = y.$$

Exercise 39: *Show that* $\mathbf{R}^n$ *is homotopy equivalent to a point. Assuming Theorem 1.44 below, conclude that*

$$H^k_{dR}(\mathbf{R}^n) \cong \begin{cases} \mathbf{R}, & k = 0, \\ 0, & k \neq 0. \end{cases}$$

To show the homotopy equivalence property, we first consider a smooth map $f : M \to N$. This induces the pullback map $f^* : \Lambda^k N \to \Lambda^k M$. The equation $f^* d\omega = df^* \omega$ yields

$$d\omega = 0 \implies df^* \omega = f^* d\omega = 0,$$

and

$$\theta = d\omega \implies f^* \theta = f^* d\omega = df^* \omega.$$

Thus f^* takes closed forms to closed forms and exact forms to exact forms. Therefore f^* induces a map, also called f^*, on de Rham cohomology,

$$f^* : H^k_{dR}(N) \to H^k_{dR}(M),$$

given by $f^*[\omega] = [f^*\omega]$.

Exercise 40: *(i) If* $g : N \to P$, *then* $(gf)^* = f^* g^* : H^k_{dR}(P) \to H^k_{dR}(M)$.
(ii) For the identity map $\mathrm{Id} : M \to M$, $\mathrm{Id}^* = \mathrm{Id} : H^k_{dR}(M) \to H^k_{dR}(M)$.

Let $i_t : M \to [0,1] \times M$ be the map $i_t(x) = (t, x)$, for any $t \in [0,1]$. Define $I : \Lambda^k([0,1] \times M) \to \Lambda^{k-1} M$ as follows: any k-form ω on $[0,1] \times M$ has the local coordinate expression

$$\begin{aligned} \omega &= f_{i_1 \ldots i_k}(t, x) dx^{i_1} \wedge \ldots \wedge dx^{i_k} + g_{j_1 \ldots j_{k-1}}(t, x) dt \wedge dx^{j_1} \wedge \ldots \wedge dx^{j_{k-1}} \\ &= \omega_1 + dt \wedge \eta, \end{aligned}$$

where $\omega_1 = f_{i_1 \ldots i_k}(t, x) dx^{i_1} \wedge \ldots \wedge dx^{i_k}$ and $\eta = g_{j_1 \ldots j_{k-1}}(t, x) dx^{j_1} \wedge \ldots \wedge dx^{j_{k-1}}$ have no dt terms. (Check that ω_1 and η are independent of choice of local coordinates on M.) Set

$$I\omega_p = \int_0^1 i_t^* \eta(t, p) dt \in \Lambda^{k-1} T_p^* M.$$

Let $d_M, d_{[0,1] \times M}$ denote the exterior derivatives on $M, [0,1] \times M$, respectively.

Lemma 1.43 $i_1^* - i_0^* = d_M \circ I + I \circ d_{[0,1] \times M}$ *on* $\Lambda^k([0,1] \times M)$.

PROOF: Both sides of the equation are linear operators on $\Lambda^k([0,1] \times M)$, so it suffices to check two cases.

1. $\omega = f_{i_1 \ldots i_k}(t, x) dx^{i_1} \wedge \ldots \wedge dx^{i_k} = f_I(t, x) dx^I$. In this case

$$d_{[0,1] \times M} \omega = (\text{garbage}) + \frac{\partial f_I}{\partial t} dt \wedge dx^I.$$

Thus

$$
\begin{aligned}
I \circ d_{[0,1] \times M} \omega &= (\int_0^1 i_t^* \frac{\partial f_I}{\partial t}\, dt) dx^I \\
&= [f_I(1, x) - f_I(0, x)] dx^I,
\end{aligned}
$$

as the reader should check, while $d_M I \omega = 0$. Now check that $i_1^* \omega - i_0^* \omega = [f(1, x) - f(0, x)] dx^I$.

2. $\eta = g_J(t, x) dt \wedge dx^J$. In this case, $i_1^* \eta = i_0^* \eta = 0$, as $i_1^* dt = i_0^* dt = 0$, and

$$
\begin{aligned}
I \circ d_{[0,1] \times M} \eta &= I(-\frac{\partial g_J}{\partial x^\alpha} dt \wedge dx^\alpha \wedge dx^J) \\
&= -(\int_0^1 \frac{\partial g_J}{\partial x^\alpha} dt) dx^\alpha \wedge dx^J, \\
d_M \circ I \omega &= d\left(\left(\int_0^1 g_J(t, x) dt\right) dx^J\right) \\
&= \frac{\partial}{\partial x^\alpha}\left(\int_0^1 g_J(t, x) dt\right) dx^\alpha \wedge dx^J.
\end{aligned}
$$

Thus $I \circ d_M \omega + d_{[0,1] \times M} \circ I \omega = 0$.

This completes the proof.

Now assume that $f_0, f_1 : M \to N$ are smoothly homotopic maps with the homotopy given by $F : [0,1] \times M \to N$, $F(0, x) = f_0(x)$, $F(1, x) = f_1(x)$. Equivalently, we have

$$
F \circ i_0 = f_0, \ f \circ i_1 = f_1.
$$

This implies

$$
\begin{aligned}
f_1^* - f_0^* &= (F \circ i_1)^* - (F \circ i_0)^* \\
&= i_1^* F^* - i_0^* F^* \\
&= dI F^* - I d F^* \quad \text{(from the lemma)} \\
&= dI F^* - I F^* d.
\end{aligned}
$$

If $[\omega] \in H^k(N)$, then

$$
f_1^* \omega - f_0^* \omega = dI F^* \omega - I F^* d\omega = dI F^* \omega,
$$

and so

$$
f_1^*[\omega] = [f_1^* \omega] = [f_0^* \omega] = f_0^*[\omega].
$$

Thus *smoothly homotopic maps of manifolds induce identical maps on de Rham cohomology groups.*

Theorem 1.44 *If M and N are smoothly homotopy equivalent manifolds, then $H_{dR}^k(M) \cong H_{dR}^k(N)$.*

PROOF. If M and N are smoothly homotopy equivalent via maps $f : M \to N$ and $g : N \to M$, then fg is homotopic to the identity map on N. By Exercise 40, the induced maps on de Rham cohomology satisfy

$$g^* f^* = (fg)^* = (\mathrm{Id}_N)^* = \mathrm{Id}_N,$$

and similarly $f^* g^* = \mathrm{Id}_M$. Thus $f^* : H^k_{dR}(N) \to H^k_{dR}(M)$ is an isomorphism of de Rham cohomology groups for all k.

Remark: The reader should be aware that there are smooth manifolds which are homeomorphic but not diffeomorphic. The first example was constructed in the 1950s by Milnor, who showed that S^7 can be given a smooth structure not diffeomorphic to the standard structure. Since then, many examples of manifolds with so-called exotic structures have been given. For example, using Donaldson's gauge theory techniques, it can be shown that $\mathbf{R}^4$ admits exotic structures, and that there exist compact four-manifolds with infinitely many distinct differentiable structures.

Our proof of the smooth homotopy invariance of the de Rham cohomology groups leaves open the possibility that these groups may depend on the choice of differentiable structure on a manifold. Fortunately, this is not the case: the de Rham cohomology groups depend only on the underlying topological structure of the manifold and are in fact (continuous) homotopy invariants. Thus these groups are topological invariants, not just invariants of differential topology.

This topological invariance is shown by de Rham's theorem, which states that the de Rham cohomology groups $H^k_{dR}(M)$ are isomorphic to $H^k_{sing}(M; \mathbf{R})$, the singular cohomology groups of M with real coefficients, which are topological invariants from their definition. For the reader familiar with singular homology and cohomology, the isomorphism of these two groups is given by the de Rham map

$$dR : H^k_{dR}(M) \to H^k_{sing}(M; \mathbf{R}), \text{ induced by } \omega \mapsto (\sigma \mapsto \int_\sigma \omega),$$

for any k-form ω and k-chain σ.

Exercise 41: *Show that the de Rham map is well defined, i.e. (i) dR is independent of the choice of representative of $[\omega]$, and (ii) the image of dR lies in the space of k-cocycles.*

Note that singular cohomology is defined for any abelian group G, and that the spaces $H^k_{sing}(M; G)$ are abelian groups. This explains why the vector spaces $H^k_{dR}(M)$ are commonly referred to as groups.

We conclude with two exercises. The solution of the second, which shows that the de Rham cohomology groups are computable for some standard spaces, depends on the Mayer-Vietoris sequence for de Rham cohomology, which can be found in [11].

Exercise 42: *Show that $H^n(M^n) \neq 0$ if M is an oriented compact manifold.*

Exercise 43: *(i) Compute the de Rham cohomology groups for S^n. Hint: Cut S^n into two overlapping pieces, each of which is diffeomorphic to the n-disk B^n, and such that the disks intersect in a cylindrical band homotopy equivalent to the equator S^{n-1}. Now use the Mayer-Vietoris sequence, induction on n, and Exercise 38.*

(ii) Show that $\beta^k(T^n) = \binom{n}{k}$. Hint: Cut T^n into two overlapping pieces, each of which is homotopy equivalent to T^{n-1}, and such that the overlap is homotopy equivalent to two copies of T^{n-1}.

(iii) Show that for a closed oriented g-holed surface Σ, $\beta^0(\Sigma) = 1$, $\beta^1(\Sigma) = 2g$, $\beta^2(\Sigma) = 1$. Hint: Cut Σ into two overlapping pieces, each of which is diffeomorphic to S_{g+1}^2, a sphere minus $g+1$ disks, and such that the overlap is homotopy equivalent to $g+1$ distinct circles. Compute the de Rham cohomology groups of S_{g+1}^2 by applying the Mayer-Vietoris sequence to S_{g+1}^2, $g+1$ disks, and their union S^2. This uses the results of Exercise 39 and (ii) above. Now use another Mayer-Vietoris sequence to compute $H_{dR}^k(\Sigma)$ from $H_{dR}^k(S_{g+1}^2)$.

1.5   The Kernel of the Laplacian on Forms

Elements of the kernel of $\Delta = \Delta^k$, the Laplacian on k-forms on a compact oriented Riemannian manifold, are called *harmonic k-forms*. Every harmonic form is closed, which gives a linear map

$$h : \operatorname{Ker} \Delta^k \to H_{dR}^k(M), \quad \omega \mapsto [\omega].$$

This leads to a second version of the Hodge theorem.

Theorem 1.45 (Hodge Theorem) *Let M be a compact oriented manifold. Then we have the isomorphism*

$$h : \operatorname{Ker} \Delta^k \xrightarrow{\cong} H_{dR}^k(M).$$

This theorem gives an important relationship between the topology of the manifold and analysis on the manifold, and has many applications, as we shall see later. Here are two immediate consequences.

Corollary 1.46 *(i) For any compact oriented manifold, $\dim H_{dR}^k(M) < \infty$ for all k.*

(ii) The dimension of the kernel of Δ^k on a compact oriented manifold is independent of the choice of Riemannian metric.

Exercise 44: *Let T^n have the metric induced from the standard metric on $\mathbf{R}^n$ as in Exercise 28. Verify Theorem 1.45 for this metric. Note that it is not feasible to verify the theorem for a general metric on the torus. Hint: Let*

$(\theta^1, \ldots, \theta^n)$ be angular coordinates on T^n. By Exercise 32, or by mimicking the calculations in §1.4, show that a basis of $H_{dR}^k(T^n)$ is given by $\{[d\theta^{i_1} \wedge \ldots \wedge d\theta^{i_k}] : i_1 < \ldots < i_k\}$. Now use the fact that the projection map $\pi : \mathbf{R}^n \to T^n$ is an isometry to show that if $a_I d\theta^I \in \operatorname{Ker} \Delta_{T^n}^k$, then

$$0 = \Delta_{\mathbf{R}^n}^k((a_I \circ \pi) dx^I) = (\Delta_{\mathbf{R}^n}^0(a_I \circ \pi)) dx^I.$$

Conclude that $a_I \in \operatorname{Ker} \Delta_{T^n}^0$, and so a_I is constant. Thus $\operatorname{Ker} \Delta_{T^n}^k$ has basis $\{d\theta^{i_1} \wedge \ldots \wedge d\theta^{i_k} : i_1 < \ldots < i_k\}$.

We will give two proofs of Theorem 1.45. The first one depends on the C^∞ theory of §3.4. The second heat equation proof, which takes place almost entirely in the L^2 framework, is longer, but has applications to the Hodge theory of noncompact manifolds.

PROOF I OF THEOREM 1.44. The map h is trivially injective. For if $[\omega] = 0 \in H_{dR}^k(M)$, then $\omega = d\theta$ for some $\theta \in C^\infty \Lambda^{k-1} T^* M$. Then

$$0 = \langle \delta\omega, \theta \rangle = \langle \delta d\theta, \theta \rangle = \langle d\theta, d\theta \rangle,$$

which implies $\omega = d\theta = 0$. For surjectivity, pick $[\alpha] \in H_{dR}^k(M)$ and a representative $\alpha \in [\alpha]$. By the Hodge decomposition theorem, we may write $\alpha = \omega + d\beta_1 + \delta\beta_2$, with ω harmonic and β_i smooth. Since $d\alpha = 0$, we have $d\delta\beta_2 = 0$, and so $0 = \langle d\delta\beta_2, \beta_2 \rangle = \langle \delta\beta_2, \delta\beta_2 \rangle$. Thus $\delta\beta_2 = 0$, and so $\alpha = \omega + d\beta_1$. Thus $\omega \in [\alpha]$.

To start the heat equation proof, recall that we are assuming that the heat operator $e^{-t\Delta}$ on k-forms has a smooth kernel $e(t, x, y) \in C^\infty(\mathbf{R} \times \Lambda^k T^* M \otimes \Lambda^k T^* M)$ as a bundle over $M \times M$. As in the case of functions, it is convenient to just write

$$(e^{-t\Delta}\alpha)(x) = \int_M e(t, x, y) \alpha(y) \, dy.$$

Lemma 1.47 $\Delta e^{-t\Delta} = e^{-t\Delta}\Delta$ and $de^{-t\Delta^k} = e^{-t\Delta^{k+1}}d$ for all smooth forms.

PROOF. We have

$$\Delta e^{-t\Delta}\omega(x) = \Delta_x \left(\int e(t, x, y)\omega(y) dy \right) = \int \Delta_x e(t, x, y)\omega(y) dy$$

$$= -\int \partial_t e(t, x, y)\omega(y) dy,$$

and, by the symmetry of $e(t, x, y)$ in x and y,

$$e^{-t\Delta}\Delta\omega(x) = \int e(t, x, y)\Delta_y\omega(y) dy = \int \Delta_y e(t, x, y)\omega(y) dy$$

$$= \int \Delta_x e(t, x, y)|_{(x,y)=(y,x)}\omega(y) dy = -\int \partial_t e(t, y, x)\omega(y) dy$$

$$= -\int \partial_t e(t, x, y)\omega(y) dy.$$

The second half of the lemma is more difficult, as the following exercise shows.

Exercise 45: *(i) Why is the following argument that $de^{-t\Delta} = e^{-t\Delta}d$ incorrect? If $\Delta^k \omega_i = \lambda_i \omega_i$, then $de^{-t\Delta^k}\omega_i = de^{-t\lambda_i}\omega_i = e^{-t\lambda_i}d\omega_i$. Since $d\Delta^k = \Delta^{k+1}d$, we have $\Delta^{k+1}d\omega_i = d\Delta^k\omega_i = d\lambda_i\omega_i = \lambda_i d\omega_i$. Thus $e^{-t\Delta^{k+1}}d\omega_i = e^{-t\lambda_i}d\omega_i$, which shows that $de^{-t\Delta} = e^{-t\Delta}d$ on the basis $\{\omega_i\}$ of $L^2\Lambda^k$.*
(ii) Use Proposition 1.38 to correct the proof in (i).

We now consider the long time behavior of the heat flow. Let $\{\omega_i\}$ be an orthonormal basis of $L^2\Lambda^k$ with $\Delta^k \omega_i = \lambda_i \omega_i$. For any $\omega \in L^2\Lambda^k$, since $e^{-t\Delta}$ is bounded on $L^2\Lambda^k$, we have

$$
\begin{aligned}
\lim_{t\to\infty} e^{-t\Delta}\omega &= \lim_{t\to\infty} e^{-t\Delta}(\sum a_i\omega_i) = \lim_{t\to\infty}\sum a_i e^{-t\Delta}\omega_i \\
&= \lim_{t\to\infty}\sum a_i e^{-\lambda_i t}\omega_i = \sum_{i=1}^{N} a_i\omega_i,
\end{aligned}
$$

where $\{\omega_1,\ldots,\omega_N\}$ is an orthonormal basis of Ker Δ. Thus a form ω flows to its harmonic component $P\omega \in$ Ker Δ, just as in the case of functions.

PROOF II OF THEOREM 1.45. Take a class $[\omega] \in H^k_{dR}(M)$. Then

$$
\begin{aligned}
e^{-t\Delta}\omega - \omega &= e^{-t\Delta}\omega - \mathrm{Id}(\omega) = \int_0^t \partial_t(e^{-t\Delta}\omega)dt \\
&= -\int_0^t \Delta e^{-t\Delta}\omega\,dt = -\int_0^t e^{-t\Delta}\Delta\omega\,dt \\
&= -\int_0^t e^{-t\Delta}d\delta\omega\,dt = -\int_0^t de^{-t\Delta}\delta\omega\,dt \\
&= d[\int_0^t e^{-t\Delta}\delta\omega\,dt],
\end{aligned}
$$

since $d\omega = 0$. This shows that the heat flow takes closed forms to closed forms, that the cohomology class of a closed form is unchanged under the heat flow, and that each cohomology class contains at least one harmonic form representative, namely $[\lim_{t\to\infty} e^{-t\Delta}\omega]$.

To finish the proof, we need only show that there is at most one harmonic form in each cohomology class. This proceeds as before. If η_1, η_2 are harmonic forms with $\eta_1 = \eta_2 + d\theta$, then $0 = \delta\eta_1 = \delta\eta_2 + \delta d\theta = \delta d\theta$. Thus $0 = \langle\theta, \delta d\theta\rangle = \langle d\theta, d\theta\rangle$, and so $d\theta = 0$. This forces $\eta_1 = \eta_2$.

Remark: By spectral theory for unbounded operators, it follows from $d\Delta = \Delta d$ that $de^{-t\Delta} = e^{-t\Delta}d$, which avoids using Proposition 1.38. Thus there is a heat equation proof of Theorem 1.45 that uses C^∞ regularity theory only to show that harmonic forms are smooth (as this ensures that the map h is well defined).

Here are some easy applications of our latest version of the Hodge theorem.

Corollary 1.48 *Let M be an n-dimensional compact, connected, oriented manifold. Then $H_{dR}^k(M) \cong H_{dR}^{n-k}(M)$. In particular,* dim $H_{dR}^n(M) = 1$.

PROOF. Since $*\Delta^k = \Delta^{n-k}*$, the map $\omega \mapsto *\omega$ is an isomorphism of Ker Δ^k and Ker Δ^{n-k}.

Exercise 46: *Let g_1, g_2 be two Riemannian metrics on a compact, connected, oriented manifold M with* vol$(M, g_1) =$ vol(M, g_2). *Show that there exists an $(n-1)$-form θ such that* dvol$(g_1) = d\theta +$ dvol(g_2). *Hint: Show that the map $\int : H_{dR}^n(M) \to \mathbf{R}$ given by $[\omega] \mapsto \int_M \omega$ is well defined and an isomorphism.*

The *Euler characteristic* of a compact manifold M is defined to be

$$\chi(M) = \sum_i (-1)^i \beta^i = \sum_i (-1)^i \text{dim Ker } \Delta^i.$$

By the de Rham isomorphism, $\chi(M) = \sum_i (-1)^i$ dim $H_{sing}^i(M; \mathbf{R})$. The reader may be familiar with the definition of the Euler characteristic given in terms of a triangulation of M: if there are b_k k-simplices in the triangulation, then $\chi(M) = \sum_i (-1)^i b_i$. In algebraic topology courses, it is shown that the singular theory and the simplicial theory produce the same cohomology groups, so $\chi(M) = \sum_i (-1)^i$ dim $H_{simp}^i(M; \mathbf{R})$. An easy argument then gives

$$\sum_i (-1)^i \text{ dim } H_{simp}^i(M; \mathbf{R}) = \sum_i (-1)^i b_i.$$

This shows that all the definitions of $\chi(M)$ agree. There are several other equivalent definitions of the Euler characteristic of a smooth manifold [32], as the Euler characteristic is in many respects the simplest topological invariant of a smooth manifold.

We now give an analytic proof of the following topological result.

Theorem 1.49 *Let M be an odd dimensional compact manifold. Then $\chi(M) = 0$.*

PROOF. If M is oriented, introduce a Riemannian metric on M and apply the last corollary. If M is nonorientable, it has a connected orientable double cover M'. Triangulate M and lift the triangulation to the double cover. The number of k-simplices doubles for each k, which shows that $\chi(M) = \chi(M')/2 = 0$.

In this last proof, we have shown that if $\pi : M' \to M$ is a finite covering map with ℓ sheets, then $\chi(M') = \ell\chi(M)$. The Betti numbers do not have this multiplicativity in general, as one can see by considering $H_{dR}^1(S^1)$. Nevertheless, we can still get some information from Hodge theory.

Theorem 1.50 *Let M' be a finite cover of a compact oriented manifold M. Then $\beta^k(M') \geq \beta^k(M)$.*

PROOF. Let $\pi : M' \to M$ be the projection map. Pick a metric g on M and pull it back to $\pi^* g$ on M'. (By definition, $\pi^* g(X, Y) = g(\pi_* X, \pi_* Y)$.) Informally speaking, the manifold M and its cover M' are locally topologically indistinguishable, and with the pullback metric they are locally geometrically indistinguishable. In particular, the Laplacian looks the same on the manifold and its cover: $\Delta_{M'} \pi^* \omega = \pi^* \Delta_M \omega$ for all forms ω on M. Thus the map $\omega \mapsto \pi^* \omega$ takes Ker Δ_M^k to Ker $\Delta_{M'}^k$, and the map is easily seen to be injective.

It is instructive to prove this last result directly using de Rham cohomology. We have $\pi_* : H_{dR}(M) \to H_{dR}(M')$ given by $[\omega] \to [\pi^* \omega]$. To determine whether this map is injective, we must show that if $\pi^* \omega = d\theta'$ on M', then there exists a form θ on M with $\omega = d\theta$. Note that θ' does not define a form on M in general, since it need not be invariant under the deck transformations of the covering map (cf. the computation of $H_{dR}^1(T^2)$). However, we can produce a form $\pi_* \alpha$ on M from any form α on M' by averaging over the fiber $\pi^{-1}(x)$:

$$\pi_* \alpha_x(X_1, \ldots, X_k) = \sum_{\tilde{x} \in \pi^{-1}(x)} \alpha_{\tilde{x}}(\pi_*^{-1} X_1, \ldots, \pi_*^{-1} X_k).$$

It is obvious that d commutes with averaging, and that $\pi_* \pi^* \omega = \omega$. Thus $\pi^* \omega = d\theta'$ implies $\omega = d\pi_* \theta'$. Hopefully, this second proof will convince the reader that Hodge theoretic proofs can be easier than de Rham cohomology arguments.

Exercise 47: *Prove that*

$$\beta^k(M \times N) = \sum_{i=0}^{k} \beta^i(M) \beta^{k-i}(N).$$

Conclude that $\chi(M \times N) = \chi(M)\chi(N)$. Show that $\beta^k(T^n) = \binom{n}{k}$ (cf. Exercise 43(ii)). Hint: Pick Riemannian metrics g, h on M, N, respectively, and form the product metric $g \oplus h$ on $M \times N$: i.e. write $X \in T_{(m,n)}(M \times N) \cong T_m M \oplus T_n N$ as $X = X_M + X_N$, and similarly for another vector Y, and set $(g \oplus h)(X, Y) = g(X_M, Y_M) + h(X_N, Y_N)$. Show that $\mathrm{dvol}_{M \times N} = \mathrm{dvol}_M \wedge \mathrm{dvol}_N$, and that $L^2(M \times N) \cong L^2(M) \hat{\otimes} L^2(N)$, where the completed tensor product symbol $\hat{\otimes}$ indicates that a Hilbert space basis of $L^2(M \times N)$ is given by $\{f_i g_j\}$ where $\{f_i\}, \{g_j\}$ are bases of $L^2(M), L^2(N)$, respectively. Generalize this to show that

$$L^2 \Lambda^k(M \times N) \cong \bigoplus_{i=0}^{k} L^2 \Lambda^i(M) \hat{\otimes} L^2 \Lambda^{k-i}(N). \qquad (1.51)$$

Show that with respect to the decomposition (1.51), the Laplacian Δ^k on $M \times N$ for the product metric takes the form

$$\Delta^k = \sum_i (\Delta_M^i \otimes \mathrm{Id} + \mathrm{Id} \otimes \Delta_N^{k-i}).$$

Conclude that a basis of the harmonic k-forms on $M \times N$ with respect to the product metric is given by $\{\alpha_j^i \otimes \beta_\ell^{k-i} : i = 0, \ldots, k\}$, where $\{\alpha_j^i\}$ is a basis of Ker Δ_M^i and $\{\beta_\ell^{k-i}\}$ is a basis of Ker Δ_N^{k-i}.

Exercise 48: *Let $S^3 * S^4$ be the join of S^3 and S^4 given by connecting every point in S^3 to every point in S^4 by a unit segment. In other words, $S^3 * S^4$ equals $S^3 \times [0, 1] \times S^4$ modulo the relations*

$$(x_1, 0, y_1) \sim (x_1, 0, y_2), \quad (x_1, 1, y_1) \sim (x_2, 1, y_1),$$

for all $x_1, x_2 \in S^3, y_1, y_2 \in S^4$. Compute the de Rham cohomology groups of $S^3 * S^4$. Do the same for $S^3 * T^4$. Hint: Use a Mayer-Vietoris sequence on the pieces $S^3 \times [0, 2/3) \times S^4$ and $S^3 \times (1/3, 1] \times S^4$. You'll need the results of Exercise 47.

Chapter 2

Elements of Differential Geometry

In the first chapter we discussed heat flow on a compact manifold and the topological significance of the long time behavior of the heat flow. In contrast, the short time behavior of the heat flow might appear trivial, as we know the heat operator goes to the identity operator as $t \downarrow 0$. However, we shall see in Chapter 3 that the way in which the heat kernel approaches the delta function (the kernel of the identity operator) is determined by the local Riemannian geometry of the manifold.

This chapter covers those parts of Riemannian geometry used to construct the heat kernel and its short time asymptotics in Chapter 3. We also prove Gårding's inequality from Chapter 1, and develop some of the supersymmetric techniques used to prove the Chern-Gauss-Bonnet theorem in Chapter 4. The key concepts discussed are the various types of curvature in Riemannian geometry (§2.1), the Levi-Civita connection associated to a Riemannian metric (§2.2.1), the Weitzenböck formula and Gårding's inequality (§2.2.2), geodesics and Riemannian normal coordinates (§2.3). There is a technical section on the Laplacian in normal coordinates (§2.4). Other references for this material include [4], [27], [64, Vols. I, II].

2.1 Curvature

There is no better place to begin a discussion of curvature than with Gauss' solution to the question: when is a piece of a surface in $\mathbf{R}^3$ (such as the earth's surface) flat? By flat, we mean that there should exist a distortion free – i.e. isometric – map from the piece of the surface to a region in the standard plane. Thus a region is flat if it possesses an accurate map in the ordinary sense. Our intuition is that this should be possible iff the region is not curved in some sense. Gauss' work gives a precise meaning to this notion of curvature, and proves that it is the obstruction to solving the mapping problem.

Locally, the given surface M is oriented and hence has an outward pointing unit vector n_x at each point $x \in M$. Consider the Gauss map $\nu : M \to$
S^2, $\nu(x) = n_x$. As test cases, consider the family of spheres $S^2(r)$ of radius r.
Notice that for $\epsilon \ll r$, the image $\nu(A)$ of the set A of points on $S^2(r)$ of distance
at most ϵ from the north pole has area proportional to r^{-2}. Moreover, for M the
standard plane, the area of the image of $\nu(A)$ is zero, where we replace the north
pole by any point in the plane. Thus the more curved the surface, the greater
the area of $\nu(A)$. It is important to note that if M is a standard cylinder, then
$\nu(A)$ still has area zero. This may seem to give the counterintuitive impression
that the cylinder is as flat as the plane, but it is in agreement with the fact that
there exists an accurate map of the cylinder to the plane given by unrolling the
cylinder; alternatively, the fact that the area is zero for the cylinder agrees with
our habit of storing maps by rolling them into cylinders.

For our test cases, there is no need to specify the radius ϵ of A, as it is easy
to check that the ratio area$(\nu(A))/$area(A) is constant. However, for arbitrary
bumpy surfaces this ratio will certainly depend on the choice of A, and a sphere
which is very bumpy very close to the north pole and very flat elsewhere will
have ratios very close to that of the flat plane unless A is very small. Thus we
would like to define the curvature at the point x to be some sort of limit

$$\text{`` } \lim_{A \to x} \frac{\text{area}(\nu(A))}{\text{area}(A)} \text{ ''.}$$

To give a precise definition, we recall that the areas above are defined as limits
of Riemann sums of areas of little boxes in tangent spaces to M and S^2, and
that det $d\nu_x$ equals

$$\frac{\text{area}(d\nu_x(B))}{\text{area}(B)},$$

for B a small box in $T_x A$. (Here it is necessary to note that $T_x M$ and $T_{\nu(x)} S^2$ are
parallel, as both are perpendicular to n_x. Thus we can canonically identify these
two tangent spaces, and so the map $d\nu_x : T_x M \to T_{\nu(x)} S^2$ has a well defined
determinant.) In summary, following Gauss we define the Gaussian curvature
as follows:

Definition: *The Gaussian curvature K_x of a surface M in $\mathbf{R}^3$ at a point x is
given by* det $d\nu_x$.

Exercise 1: *Compute the curvature of the standard two-sphere of radius r.*

Note that at the saddle point of a hyperboloid, the determinant of the Gauss
map is negative, reflecting the fact that the Gauss map is orientation reversing.

Exercise 2: *Consider the torus with its standard embedding in $\mathbf{R}^3$. Draw the
regions on the torus consisting of points of positive curvature. Do the same for
the points of negative and zero curvature.*

At this point it is reasonable to conjecture that a surface has a distortion free map ϕ from a neighborhood V of a point x to a region $U \subset \mathbf{R}^2$ iff $K_y = 0$ for all y in a neighborhood of x. Let $\{x^1, x^2\}$ be coordinates on V, with the metric g expressed as $g_{ij} dx^i \otimes dx^j$, and let $\{y^1, y^2\}$ be the standard coordinates on U. Via the map ϕ, we can as usual treat $\{y^i\}$ as coordinates on V. Then the condition that ϕ be distortion free is that the inner product in V is the same as the standard Euclidean inner product when measured with respect to y coordinates, or in other words

$$g_{ij} dx^i \otimes dx^j = dy^1 \otimes dy^1 + dy^2 \otimes dy^2. \tag{2.1}$$

Thus we want to find the functions $y^i = y^i(x^1, x^2)$, $i = 1, 2$, satisfying (2.1). If we write $dy^i = (\partial y^i / \partial x^j) dx^j$, we see that (2.1) is a system of three nonlinear first order partial differential equations for the unknown functions y^i. Moreover, we expect the system to have a solution iff the Gaussian curvature vanishes on V. As we will now explain, in the language of PDEs this means that we expect the Gaussian curvature to be the integrability condition for this system.

As a simple example of an integrability condition, consider the following problem in classical vector calculus:

Example: *Given a vector field $X = a(x, y)\partial_x + b(x, y)\partial_y$ on $\mathbf{R}^2$, find a function f with $\nabla f = X$.*

Of course, we must solve

$$\frac{\partial f}{\partial x} = a, \quad \frac{\partial f}{\partial y} = b.$$

This is not always possible, since we must have

$$\frac{\partial a}{\partial y} = \frac{\partial^2 f}{\partial x \partial y} = \frac{\partial b}{\partial x}. \tag{2.2}$$

It is a basic result that (2.2) is a necessary and sufficient condition for solving $\nabla f = X$, and that the solution is given by a standard integration technique. Thus the equation

$$\frac{\partial a}{\partial y} = \frac{\partial b}{\partial x}$$

is called the integrability condition of the equation $\nabla f = X$. (We have already seen this condition in §1.4.)

In this simple example, the integrability condition was fairly obvious to discover. Note, however, that writing any mixed partial derivative of f in terms of a and b gives another condition that X must satisfy. The only real work in solving $\nabla f = X$ is in determining which mixed partial condition is sufficient for solving the equation.

For a general system of PDEs, even the first step of finding any restrictions on the given data may be difficult, as the equations that result from setting mixed partials equal may still contain unknown functions and not just the given data; the reader is urged to try this for the system (2.1). Thus finding integrability conditions is not just a mechanical process.

In any case, our conjecture for the solution of the mapping problem should be reinterpreted as the conjecture that the vanishing of the Gaussian curvature is the integrability condition for (2.1). In fact, Gauss found the integrability condition for the system (2.1): we must have

$$\frac{\partial \Gamma_{22}^1}{\partial x^1} - \frac{\partial \Gamma_{21}^1}{\partial x^2} + \Gamma_{22}^s \Gamma_{s1}^1 - \Gamma_{21}^s \Gamma_{s2}^1 = 0 \qquad (2.3)$$

in a neighborhood of x, where the *Christoffel symbols* Γ_{jk}^i are defined by

$$\Gamma_{jk}^i = \frac{1}{2} g^{il} \left(\frac{\partial g_{kl}}{\partial x^j} + \frac{\partial g_{jl}}{\partial x^k} - \frac{\partial g_{jk}}{\partial x^l} \right).$$

Recall from Exercise 14, Chapter 1, that $(g^{kl}) = (g_{ij})^{-1}$.

Although Gauss has given us the analytic solution to the mapping problem, it is not obvious how it is related to our geometric intuition that the solution be determined by the Gaussian curvature. In fact, at first glance it seems implausible that the two approaches should be equivalent, since the Gaussian curvature is defined in terms of the surface's placement in $\mathbf{R}^3$, while the analytic solution (2.3) is defined strictly in terms of the geometry of the surface. In other words, if $f : M \to N$ is a map of surfaces in $\mathbf{R}^3$ which is an isometry with respect to the induced metrics, then (2.3) holds for M iff it holds for N. It is not clear at all that this property is true for the Gaussian curvature, since f need not be the restriction of an isometry of all of $\mathbf{R}^3$. (For example, f might be the map from the plane to the cylinder which rolls up the plane.) Nevertheless, Gauss proved his intuition correct by showing that the Gaussian curvature is indeed given by the left hand side of (2.3) divided by $|dx^1 \wedge dx^2|^2$, the square of the area of the parallelogram in $T_x^* M$ spanned by dx^1, dx^2. Thus the curvature can be defined directly off the metric tensor of the surface. This is Gauss' famous Theorema Egregium, dating from around 1830, which we state in the following form:

Theorem 2.4 *A surface in $\mathbf{R}^3$ admits a distortion free map from a neighborhood of a point to a region in the flat plane iff the left hand side of (2.3) vanishes in a neighborhood of the point. Moreover, the left hand side of (2.3) divided by $|dx^1 \wedge dx^2|^2$ equals the Gaussian curvature of the surface.*

In particular, there are no distortion free maps of regions on the standard sphere to the plane.

We will prove a more general form of the first part of the theorem in Theorem 2.10. Moreover, since the left hand side of (2.3) is defined for any Riemannian two-manifold, not necessarily embedded in $\mathbf{R}^3$, we can *define* the Gaussian curvature K of an arbitrary Riemannian surface to be this left hand side divided by

$|dx^1 \wedge dx^2|^2$. Since this quantity equals $\det(g)$ computed in (x^1, x^2) coordinates (as the reader should check), we have by definition

$$K = \frac{1}{\det(g)} \left(\frac{\partial \Gamma_{22}^1}{\partial x^1} - \frac{\partial \Gamma_{21}^1}{\partial x^2} + \Gamma_{22}^s \Gamma_{s1}^1 - \Gamma_{21}^s \Gamma_{s2}^1 \right). \qquad (2.5)$$

While the Gaussian curvature is the solution to a local problem, it has strong influence on the global topology of a surface. Recall that the Euler characteristic of the g-holed torus is $2 - 2g$, as can be computed from a triangulation of the surface.

Theorem 2.6 (Gauss-Bonnet Theorem) *If M is a closed surface in $\mathbf{R}^3$, then*

$$\chi(M) = \frac{1}{2\pi} \int_M K \, dA.$$

Here dA is the classical notation for the volume element on M in the induced metric. It should seem remarkable that the integral is independent of any deformation of the surface in $\mathbf{R}^3$. The Gauss-Bonnet theorem remains true for arbitrary Riemannian surfaces, with the Gaussian curvature defined by (2.5). As we will see, the Gauss-Bonnet theorem (ca. 1850) is historically the first instance of the Atiyah-Singer index theorem.

The Gauss-Bonnet theorem shows that the global topology of a surface restricts the type of Riemannian metrics the surface admits:

Corollary 2.7 *(i) S^2 has no metric with $K \leq 0$ everywhere.*
(ii) T^2 has no metric with $K > 0$ or $K < 0$ everywhere.
(iii) An oriented closed surface of genus $g > 1$ has no metric with $K \geq 0$ everywhere.

The Gauss-Bonnet theorem gives the only topological obstruction to the curvature function on a closed surface. That is, by Exercise 1 the sphere admits a metric with $K \geq 0$. Furthermore:

Exercise 3: *(i) T^2 admits a metric with $K \equiv 0$. Hint: use Exercise 28, Chapter 1.*
(ii) Consider the upper half plane $\mathcal{H} = \{(x, y) : y > 0\}$ with the metric tensor $y^{-2}(dx \otimes dx + dy \otimes dy)$. Show that the curvature of $\mathcal{H}$ is identically -1.

It is a nontrivial fact that there exist subgroups Γ_g of the group of isometries of $\mathcal{H}$ such that the quotient manifold $\mathcal{H}/\Gamma_g$ is a closed surface with g holes, for any $g > 1$. As in Exercise 28, Chapter 1, this implies that these surfaces have a metric of constant negative curvature. It is interesting to note that (more or less) explicit expressions for the harmonic one-forms for these constant curvature metrics were not known until fairly recently [40].

Around 1860, Riemann gave an inaugural lecture for an unpaid lectureship, in which he demonstrated a clear understanding of manifolds with Riemannian metric (without giving a precise definition of either by modern standards). Within a few years, he had posed and solved the basic question of when a piece of a Riemannian manifold admits a distortion free map to $\mathbf{R}^n$ with the standard metric. In other words, Riemann found the integrability condition for the system

$$g_{ij}dx^i \otimes dx^j = \delta_{ij}dy^i \otimes dy^j, \tag{2.8}$$

the natural generalization of (2.1). To set the notation, we define the *Riemann curvature tensor* of type (1,3) to be

$$R = R^i_{jkl}\partial_{x^i} \otimes dx^j \otimes dx^k \otimes dx^l \in \mathcal{T}^3_1,$$

where

$$R^i_{jkl} = \frac{\partial \Gamma^i_{jk}}{\partial x^l} - \frac{\partial \Gamma^i_{jl}}{\partial x^k} + \Gamma^s_{jk}\Gamma^i_{sl} - \Gamma^s_{jl}\Gamma^i_{sk}, \tag{2.9}$$

and the Γ^i_{jk} are defined as before. It is not obvious that R^i_{jkl} are the components of a tensor, although a long calculation verifies that the R^i_{jkl} do transform correctly; see Exercise 4 below. The notation $R \in \mathcal{T}^3_1$ means that R is a smooth section of $TM \otimes T^*M \otimes T^*M \otimes T^*M$. (Warning: while (2.9) agrees with the definition in [27], other texts such as [64] define the curvature tensor to be the negative of our definition.)

Theorem 2.10 (Riemann) *The system (2.8) has a solution in the neighborhood of a point iff $R \equiv 0$ in a (possibly different) neighborhood of the point.*

A metric with $R \equiv 0$ is called *flat*.

The proof is a beautiful example of 19th century PDE techniques. We will show that a solution exists only if the curvature tensor vanishes; a complete proof is in [64, Vol. II]. Throughout the messy calculation, keep in mind that we are trying to eliminate the unknown functions y^i to obtain an integrability condition. The only techniques available for the proof are differentiation of the system (2.8) and algebraic manipulations.

PROOF. Given coordinates $(x^1, \ldots, x^n)$ with the associated matrix (g_{ij}) of inner products, we assume that there are functions $y^1 = y^1(x^1, \ldots, x^n), \ldots, y^n = y^n(x^1, \ldots, x^n)$ such that $(y^1, \ldots, y^n)$ are also local coordinates on the manifold which solve (2.8). Plugging $dy^i = (\partial y^i / \partial x^k)dx^k, dy^j = (\partial y^j / \partial x^l)dx^l$ into (2.8), we see that solving (2.8) is equivalent to finding functions $y^1, \ldots, y^n$ such that

$$g_{ij} = \frac{\partial y^\alpha}{\partial x^i}\frac{\partial y^\alpha}{\partial x^j}. \tag{2.11}$$

(Here and below we abuse the Einstein summation convention by summing over repeated superscripts.) Let $A = \left(\dfrac{\partial y^\alpha}{\partial x^i}\right)$ be the Jacobian matrix for the y maps, and let $G = (g_{ij})$ be the matrix of the metric with inverse matrix entries denoted $G^{-1} = (g^{ij})$. Then (2.11) is equivalent to $G = A^T A$, or $AG^{-1}A^T = \text{Id}$. (A^{-1} exists because the map taking a point's x coordinates to its y coordinates is a diffeomorphism of $\mathbf{R}^n$.)

Letting δ^μ_ν denote the Kronecker delta symbol, we see that the equation $AG^{-1}A^T = \text{Id}$ and hence (2.11) are equivalent to

$$\frac{\partial y^\mu}{\partial x^i} g^{ij} \frac{\partial y^\nu}{\partial x^j} = \delta^\mu_\nu. \tag{2.12}$$

Differentiating (2.11) gives

$$\frac{\partial g_{ij}}{\partial x^k} = \frac{\partial y^\alpha}{\partial x^j} \frac{\partial^2 y^\alpha}{\partial x^i \partial x^k} + \frac{\partial y^\alpha}{\partial x^i} \frac{\partial^2 y^\alpha}{\partial x^j \partial x^k}.$$

There are similar expressions for $\partial g_{ik}/\partial x^j, \partial g_{jk}/\partial x^i$. Adding these three equations with appropriate signs gives

$$\frac{1}{2}\left(\frac{\partial g_{ij}}{\partial x^k} + \frac{\partial g_{ik}}{\partial x^j} - \frac{\partial g_{jk}}{\partial x^i}\right) = \frac{\partial^2 y^\alpha}{\partial x^j \partial x^k} \frac{\partial y^\alpha}{\partial x^i}.$$

Now it is time for an algebraic trick. Fix a new index λ. The last equation gives

$$\begin{aligned}\frac{1}{2}g^{i\gamma}\frac{\partial y^\lambda}{\partial x^\gamma}\left(\frac{\partial g_{ij}}{\partial x^k} + \frac{\partial g_{ik}}{\partial x^j} - \frac{\partial g_{jk}}{\partial x^i}\right) &= g^{i\gamma}\frac{\partial y^\lambda}{\partial x^\gamma}\frac{\partial^2 y^\alpha}{\partial x^j \partial x^k}\frac{\partial y^\alpha}{\partial x^i} \\ &= \frac{\partial^2 y^\alpha}{\partial x^j \partial x^k}\delta^\alpha_\lambda. \end{aligned} \tag{2.13}$$

Here we have used (2.12) with the index of summation j replaced by γ. Notice that the left hand side of (2.13) is just $\Gamma^\gamma_{jk}\frac{\partial y^\lambda}{\partial x^\gamma}$. Thus (2.13) becomes

$$\frac{\partial^2 y^\lambda}{\partial x^j \partial x^k} = \Gamma^\gamma_{jk}\frac{\partial y^\lambda}{\partial x^\gamma}. \tag{2.14}$$

This equation is actually a large improvement over the original system (2.8), as we have decoupled this nonlinear system into a linear system of equations, each of which involves only one y^λ. In fact, the new system involves only first and second partials of y^λ, so we can do a standard reduction of order. For fixed λ, define a vector of functions

$$z = (z^1, \ldots, z^n) = \left(\frac{\partial y^\lambda}{\partial x^1}, \ldots, \frac{\partial y^\lambda}{\partial x^n}\right).$$

Then (2.14) is just

$$\frac{\partial z^j}{\partial x^k} = \Gamma^\gamma_{jk}z^\gamma. \tag{2.15}$$

Since we are assuming that a solution to (2.15) exists, the mixed partials of the z functions must be equal:

$$\frac{\partial^2 z^j}{\partial x^\ell \partial x^k} = \frac{\partial^2 z^j}{\partial x^k \partial x^\ell}.$$

Looking at (2.15), we see we must have

$$\frac{\partial}{\partial x^\ell} \Gamma^\gamma_{jk} z^\gamma = \frac{\partial}{\partial x^k} \Gamma^\gamma_{j\ell} z^\gamma$$

for all $j, k, \ell \in \{1, \ldots, n\}$. Equivalently, if we set $f^j_k(x, z) = f^j_k(x, z(x)) = \Gamma^j_{jk}(x) z^\gamma(x)$, we must have

$$\frac{\partial f^j_k}{\partial x^\ell} = \frac{\partial f^j_\ell}{\partial x^k}.$$

By the chain rule, this becomes

$$\frac{\partial f^j_k}{\partial x^\ell} + \frac{\partial f^j_k}{\partial z^\mu}\frac{\partial z^\mu}{\partial x^\ell} - \frac{\partial f^j_\ell}{\partial x^k} - \frac{\partial f^j_\ell}{\partial z^\mu}\frac{\partial z^\mu}{\partial x^k} = 0,$$

or

$$\frac{\partial}{\partial x^\ell}(\Gamma^\gamma_{jk} z^\gamma) + \frac{\partial}{\partial z^\mu}(\Gamma^\beta_{jk} z^\beta)\Gamma^\gamma_{\mu\ell} z^\gamma - \frac{\partial}{\partial x^k}(\Gamma^\gamma_{j\ell} z^\gamma) - \frac{\partial}{\partial z^\mu}(\Gamma^\beta_{j\ell} z^\beta)\Gamma^\gamma_{\mu k} z^\gamma = 0.$$

Since $\partial z^\beta / \partial z^\mu = \delta^\beta_\mu$, we get

$$\frac{\partial}{\partial x^\ell}(\Gamma^\gamma_{jk}) z^\gamma + \Gamma^\mu_{jk}\Gamma^\gamma_{\mu\ell} z^\gamma - \frac{\partial}{\partial x^k}(\Gamma^\gamma_{j\ell}) z^\gamma - \Gamma^\mu_{j\ell}\Gamma^\gamma_{\mu k} z^\gamma = 0.$$

Now $z = (z^1, \ldots, z^\gamma, \ldots, z^n)$ in (2.15) has an arbitrary initial value, also denoted by z, at some chosen initial point x, and the last equation expresses that z is orthogonal to the vector with components

$$\frac{\partial}{\partial x^\ell}\Gamma^\gamma_{jk} + \Gamma^\mu_{jk}\Gamma^\gamma_{\mu\ell} - \frac{\partial}{\partial x^k}\Gamma^\gamma_{j\ell} - \Gamma^\mu_{j\ell}\Gamma^\gamma_{\mu k}. \tag{2.16}$$

This forces (2.16) to vanish. Since (2.16) is precisely R^γ_{jkl}, the curvature tensor must vanish to solve the mapping problem.

Exercise 4: *The condition $R^i_{jk\ell} = 0$ must be independent of the choice of local coordinates, as it is the integrability condition for the mapping problem, whose statement does not depend on a choice of coordinates. This indicates that the $R^i_{jk\ell}$ should be the components of a tensor. Show that $R^i_{jk\ell}$ are the components of a tensor of type (1,3) (i.e. $R^i_{jkl}\partial_{x^i} \otimes dx^j \otimes dx^k \otimes dx^l$ is a tensor). Hint: Let $(y^1, \ldots, y^n)$ be another set of coordinates, and write*

$$g_{ij}dx^i \otimes dx^j = h_{rs}dy^r \otimes dy^s, \tag{2.17}$$

where $h_{rs} = \langle \partial_{y^r}, \partial_{y^s} \rangle$. Recall that g_{ij} and h_{rs} are related as in Exercise 8, Chapter 1. Now perform all the manipulations we did to obtain $R^i_{jk\ell}$ from the *various g_{ij}, and do the same for the right hand side of (2.17). Then convert all the x differentiations of the right hand side to y differentiations using the chain rule $\partial_{x^i} = (\partial y^j/\partial x^i)\partial_{y^j}$. You should end up with the desired transformation law expressing $R^i_{jk\ell}$ in the x coordinates in terms of the $R^i_{jk\ell}$ in the y coordinates.*

To compare Riemann's result with Gauss' theorem requires some familiarity with raising and lowering indices, the process of creating new tensors from a given tensor via the Riemannian metric. We let $\mathcal{T}^l_k$ denote the space of tensors of type (k, l) – i.e. with k ∂_{x^i}'s and l dx^j's.

Exercise 5: *(i) If $X = a^i\partial_{x^i} \in \mathcal{T}^0_1$ is a vector field, show that $\omega = g_{ij}a^j dx^i \in \mathcal{T}^1_0$ is a one-form – i.e. show that $g_{ij}a^j dx^i \in \mathcal{T}^1_0$ is independent of the choice of local coordinates. We denote $g_{ij}a^j$ by a_i, so $\omega = a_i dx^i$. Check that $\alpha_g(X) = \omega$ in the notation of §1.2.2. Similarly, if $\omega = a_i dx^i \in \mathcal{T}^1_0$, show that $X = a^i\partial_{x^i} \in \mathcal{T}^0_1$, where $a^i = g^{ij}a_j$.*

(ii) If $A^i_{jkl}\partial_{x^i} \otimes dx^j \otimes dx^k \otimes dx^l \in \mathcal{T}^3_1$ is any tensor, then $g_{im}A^m_{jkl}dx^i \otimes dx^j \otimes dx^k \otimes dx^l$ is a tensor of type $(0, 4)$. We denote $g_{im}A^m_{jkl}$ by A_{ijkl}.

By iterating this process, we can clearly raise or lower an arbitrary number of indices in a tensor, which is equivalent to repeated applications of the map α (or α^{-1}) of §1.2.2.

Applying the exercise to the Riemann curvature tensor, it is easy to check that

$$R_{ijk\ell} = -R_{jik\ell}, \; R_{ijk\ell} = -R_{ij\ell k}, \; R_{ijk\ell} = R_{k\ell ij}. \qquad (2.18)$$

As a result, for surfaces the only nonzero component of R is R^1_{212} (and its permutations given by the last equation), and from (2.5), (2.9) we have

$$-R^1_{212} = K|dx^1 \wedge dx^2|^2. \qquad (2.19)$$

Thus Riemann's theorem generalizes Gauss' result for surfaces, as it must.

Exercise 6: *Derive the* Bianchi identity

$$R_{ijk\ell} + R_{i\ell jk} + R_{ik\ell j} = 0. \qquad (2.20)$$

Riemann's calculation can be generalized to show that two Riemannian manifolds are locally isometric iff the curvature tensors agree in a certain technical sense. Thus the Riemann curvature tensor completely determines the Riemannian geometry locally. However, in practice this is not of much use; after taking account of the symmetries of R, we still have $n(n-1)/2$ independent components for the curvature tensor on an n-manifold. For n of any reasonable size, the tensor R is just too complicated to be of much use. Therefore, it is natural to use the metric to "average" the curvature tensor.

The first simplification is the *Ricci curvature*

$$\text{Ric} = R_{ik} dx^i \otimes dx^k \in T_0^2,$$

where

$$R_{ik} = R^j_{ikj} = g^{jm} R_{imkj}.$$

The Ricci tensor is of the same type as the metric tensor, so we can compare them as inner products. Namely, we write $\text{Ric} \geq C$ if $\text{Ric}(\omega, \omega) \geq C g(\omega, \omega)$ for all $\omega \in T_x M$, for all $x \in M$. The Ricci tensor, while weaker than the curvature tensor, still controls some crucial rough geometric properties of a Riemannian manifold. For example, the volume of geodesic balls (i.e. the set of points within a fixed distance of a given point – see Exercise 9, Chapter 1) is controlled by the Ricci tensor. Let $B_r(x)$ denote the set of points within distance r of $x \in M$.

Theorem 2.21 *If* $\text{Ric} \geq C(n-1)$, *where* $n = \dim M$, *then*

$$\text{vol}(B_r(x)) \leq e^{-Cr}, \quad C < 0,$$

$$\text{vol}(B_r(x)) \leq Cr^n, \quad C = 0.$$

A proof of this result is in [27, Ch. III. H].

The Ricci curvature also influences the topology of a manifold. For example, Myers' theorem states that a compact manifold with positive Ricci curvature has finite fundamental group [46]. In contrast, it is a recent result of Lohkamp [41] that a metric having negative Ricci curvature places no restriction on the manifold, as any manifold of dimension at least three admits a metric of negative Ricci curvature.

A still weaker curvature, the *scalar curvature*, is given by contracting the two indices of the Ricci tensor. Namely, we set

$$s = R^i_i = g^{im} R_{mi}.$$

Notice that this is a function on M.

Exercise 7: *(i) Formulate a coordinate free definition of the contraction of a tensor. In other words, given a tensor of type (k, l), show how to construct a tensor of type $(k-1, l-1)$. Check that our constructions of the Ricci and scalar curvatures use this procedure. Hint: for any vector space V, there is the natural map $V^* \otimes V \to \mathbf{R}$ given by $\lambda \otimes v \mapsto \lambda(v)$.*

(ii) Show that if V has an inner product, then from a tensor of type (k, l) on V, one can construct a tensor of type $(k, l-2)$ and a tensor of type $(k-2, l)$, by means of natural maps $V^* \otimes V^* \to \mathbf{R}$ and $V \otimes V \to \mathbf{R}$. If $A^{i_1 \cdots i_k}_{j_1 \cdots j_l}$ are the components of a tensor of type (k, l), what are the components of these new tensors?

The scalar curvature is really too weak a concept to control the geometry of the manifold, although there is a global topological obstruction (the $\hat{A}$-genus of a spin manifold) to a manifold admitting a metric of positive scalar curvature, just as the Euler characteristic can be an obstruction to a surface having a metric of positive Gaussian curvature. In fact, if the dimension of M is at least three, and $f : M \to \mathbf{R}$ is any smooth function which changes sign, then there exists a metric on M (in fact many) whose scalar curvature equals f [38]. Thus if the scalar curvature changes sign, one can conclude nothing about the topology of the manifold.

The final notion of curvature, the *sectional curvature* of a two-plane in the tangent space, comes closest to mimicking the Gaussian curvature of a surface. If $P \subset T_x M$ is a two-plane spanned by $v_1, v_2 \in T_x M$ with $|v_1| = |v_2| = 1$, then we can find coordinates (x^i) near x such that $v_1 = \partial_{x^1}, v_2 = \partial_{x^2}, \langle v_1, v_2 \rangle = 0$ at x (check). We set the sectional curvature of P to be $K(P) = -R_{1212}/|v_1 \wedge v_2|^2$, where the curvature tensor is computed in the (x^i) coordinates. (Dividing by $|v_1 \wedge v_2|^2$ makes $K(P)$ independent of the choice of basis of P.) Of course, if M is a surface, then $P = T_x M$ and $K(P)$ is the Gaussian curvature at x. It is an unpleasant exercise in linear algebra to show that the $n(n-1)/2$ components $R_{ijij}, 1 \le i < j \le n$, which give the sectional curvatures (up to the factors $|\partial_i \wedge \partial_j|^2$) of the coordinate two-planes at a point, determine all the R_{ijkl} and hence determine the full curvature tensor. Thus in principle the sectional curvatures determine the local geometry of a Riemannian manifold.

Exercise 8: *(i) Show that for the standard metric on S^n all sectional curvatures equal one. Show that $\mathrm{Ric} = (n-1)g$ and that the scalar curvature is the constant $n(n-1)$. Hint: argue by symmetry considerations that it suffices to compute the sectional curvature of just one plane at the north pole.*

(ii) Compute the Ricci curvature and the scalar curvature of a surface in terms of the Gaussian curvature.

As in the case of the Gauss-Bonnet theorem, the sectional curvatures can also influence the topology of a compact manifold. Let M be a compact manifold with sectional curvature $K = K(P)$.

Theorem 2.22 *(i)* **(Cartan-Hadamard)** *If $K \le 0$ for all two-planes, then the universal cover $\widetilde{M}$ is diffeomorphic to $\mathbf{R}^n$. In particular, $\pi_k(M) = 0$ for $k > 1$.*

(ii) **(Synge)** *If $K > 0$, and M is orientable and even dimensional, then $\pi_1(M) = 0$.*

Proofs are in [64, Vol. IV].

Nevertheless, the relationship between the sectional curvature and the topology of compact manifolds is far from understood. For example, the following questions raised by Hopf in the 1950s are still unsolved:

Hopf conjectures:
    (i) If dim $M = 2n$ and $K < 0$, then $(-1)^n \chi(M) > 0$.
    (ii) If dim $M = 2n$ and $K > 0$, then $\chi(M) > 0$.

(iii) $S^2 \times S^2$ has no metric with $K > 0$.

Conjecture (i) is true if $n = 1$ by Gauss-Bonnet, and is known for $n = 2$, for manifolds of constant negative curvature, for manifolds with K pinched close to a constant, and for certain complex manifolds. As far as I know, aside from a pinching result for (ii), and a partial result in [14] for (iii), no progress has been made on the other conjectures. It is interesting to note that Hopf raised question (i) before any examples of compact manifolds of negative curvature in dimensions above three were known [10], and well before any examples of negatively curved manifolds not admitting a metric of constant negative curvature were given [50]. This is not a recommended technique for making conjectures.

2.2 The Levi-Civita Connection and Bochner Formula

In this section we will introduce the Levi-Civita connection on a Riemannian manifold, which essentially allows us to take higher order derivatives of functions and tensors. We'll see how to relate the Levi-Civita connection to the curvature of the metric, and how this connection relates to the Laplacian on forms. As applications, we'll give a proof of Gårding's inequality (Theorem 2.46) and of Bochner's theorem, which states that a manifold with positive Ricci curvature has $H^1_{dR}(M) = 0$.

2.2.1 The Levi-Civita Connection

On a smooth manifold, the differential df encodes all the first derivative information of a function f in a coordinate free manner; equivalently, on a Riemannian manifold, the gradient ∇f encodes this information. To keep track of second derivative information, $d^2 f$ certainly won't do, and Δf is a complicated combination of second derivative information (and lower order terms). Thus we need some coordinate free way of taking the derivative of ∇f, or more generally of any vector field. The main difficulty is that a vector field lies in a different tangent space $T_x M$ for each $x \in M$; to see how these vectors are changing, we need some canonical method of comparing different tangent spaces.

If the manifold is $\mathbf{R}^n$, then all tangent spaces are canonically isomorphic, and the derivative of a vector field X in the direction $v \in T_x \mathbf{R}^n$ is just the directional derivative $D_v X(x)$, which is again a vector at x. The directional derivative's crucial properties are (i) it is linear in v, and (ii) it satisfies the Leibniz rule: for any smooth function f,

$$D_v(fX)(x) = f \cdot D_v(X)(x) + df(v) \cdot X(x).$$

Exercise 9: *Why doesn't the Lie bracket $L_Y X = [X, Y]$, where X and Y are vector fields on a fixed manifold, give the desired generalization of directional derivative?*

If our manifold is a surface M in $\mathbf{R}^3$ with the induced metric, then we can define the derivative of a vector field X on M as follows. Extend X to a vector field on a small neighborhood of M in $\mathbf{R}^3$. For $v \in T_x M$, set $\nabla_v X = P \circ D_v X$, where D is the ordinary directional derivative in $\mathbf{R}^3$ and P is the orthogonal projection of $T_x \mathbf{R}^3 \cong \mathbf{R}^3$ to $T_x M$. (Check that this definition is independent of the extension of X.) Let $\Gamma(TM) = \mathcal{T}_1^0$ denote the space of sections of TM, i.e. the space of vector fields. Then $\nabla : TM \otimes \Gamma(TM) \to \Gamma(TM)$ is linear in v and satisfies the Leibniz rule $\nabla_v(fX) = f\nabla_v X + df(v)X$. In particular, ∇ is determined in a coordinate chart by the quantities $\nabla_{\partial_j} \partial_k$, where ∂_j denotes ∂_{x^j}.

Exercise 10: *Show that for this ∇ we have $\nabla_{\partial_j} \partial_k = \Gamma^i_{jk} \partial_i$. Hint: In a coordinate chart (x, y) on M, with ∂_x, ∂_y the usual basis vectors for tangent spaces to M in the chart, let ∂_{xy} denote the vector giving the derivative of ∂_x in the direction of ∂_y. Let n be the normal vector to M. Write*

$$
\begin{aligned}
\partial_{xx} &= \Lambda^1_{11}\partial_x + \Lambda^2_{11}\partial_y + L_1 n, \\
\partial_{xy} &= \Lambda^1_{12}\partial_x + \Lambda^2_{12}\partial_y + L_2 n, \\
\partial_{yx} &= \Lambda^1_{21}\partial_x + \Lambda^2_{21}\partial_y + L_3 n, \\
\partial_{yy} &= \Lambda^1_{22}\partial_x + \Lambda^2_{22}\partial_y + L_4 n,
\end{aligned}
$$

for some constants Λ^k_{ij}. Take the inner product of these four equations with ∂_x, ∂_y, and solve the resulting equations for the Λ^k_{ij}. Conclude that $\Lambda^k_{ij} = \Gamma^k_{ij}$. Note that when we project e.g. ∂_{xx} into T_M, we just omit the terms with n in the equations above. For more details, see [21, Ch. 4].

We now carry over this construction to any Riemannian manifold.

Definition 1: The Levi-Civita Connection. *Let (M, g) be a Riemannian manifold. Define $\nabla : TM \otimes \Gamma(TM) \to \Gamma(TM)$, $(v \otimes X) \mapsto \nabla_v X$, by the conditions:*
  (i) $\nabla_{\partial_j} \partial_k = \Gamma^i_{jk} \partial_i$;
  (ii) $\nabla_{\lambda v} X = \lambda \nabla_v X$ and $\nabla_{v+w} X = \nabla_v X + \nabla_w X$;
  (iii) $\nabla_v(fX) = f\nabla_v X + df(v)X$, for all smooth $f : M \to \mathbf{R}$.

$\nabla_j = \nabla_{\partial_j}$ is also called *the covariant derivative in the jth direction*. Note that condition (i) is formulated in terms of a local coordinate chart, which makes it aesthetically less appealing than the other conditions, and forces the following exercise.

Exercise 11: *(i) Show that condition (i) in Definition 1 is independent of choice of coordinate chart. Warning: this is a long computation.*
  (ii) Alternatively, for those who prefer coordinate free formulations, show that the operator $D : \Gamma(TM \otimes TM) \to \Gamma(TM)$ defined by

$$
\begin{aligned}
2\langle D_X Y, Z \rangle = \ & X\langle Y, Z \rangle + Y\langle Z, X \rangle - Z\langle X, Y \rangle \\
& + \langle [X, Y], Z \rangle + \langle [Z, X], Y \rangle - \langle [Y, Z], X \rangle \quad (2.23)
\end{aligned}
$$

satisfies (a) $D_X Y(x)$ depends only on the value of X at x, and so D may be considered as an operator $D : TM \otimes \Gamma(TM) \to \Gamma(TM)$; and (b) D is the Levi-Civita connection. Use (2.23) to conclude that the Levi-Civita connection satisfies

$$\left. \begin{array}{r} X\langle Y, Z \rangle = \langle \nabla_X Y, Z \rangle + \langle Y, \nabla_X Z \rangle, \\[2mm] \nabla_X Y - \nabla_Y X = [X, Y]. \end{array} \right\} \qquad (2.24)$$

(In fact, the Levi-Civita connection is the unique operator from $\Gamma(TM \otimes TM)$ to $\Gamma(TM)$ satisfying (2.24), see [27].)

For our purposes, it is convenient to use a little confusing linear algebra to rewrite the Levi-Civita connection.

Exercise 12: *Let V be a finite dimensional vector space and W any vector space. Show that a homomorphism $h : V \otimes W \to W$ induces a homomorphism $h' : W \to V^* \otimes W$ by $h'(w)(v) = h(v \otimes w)$.*

Applying this exercise to the Levi-Civita connection, we obtain a map ∇' : $\Gamma(TM) \to T^* M \otimes \Gamma(TM)$ given by $\nabla'(X)(v) = \nabla_v X$. In terms of a coordinate chart, this becomes $\nabla' \partial_k = dx^j \otimes \Gamma^i_{jk} \partial_i$. From now on we'll just write ∇ for ∇'. The reader should check that the following definition is equivalent to the previous one.

Definition 2: The Levi-Civita Connection. *Let (M, g) be a Riemannian manifold. Define $\nabla : \Gamma(TM) \to T^* M \otimes \Gamma(TM)$, by the conditions:*
  (i) $\nabla \partial_k = dx^j \otimes \Gamma^i_{jk} \partial_i$;
  (ii) $\nabla X(\lambda v) = \lambda \nabla X(v)$ and $\nabla X(v + w) = \nabla X(v) + \nabla X(w)$ in $\Gamma(TM)$, for all $X \in \Gamma(TM), v, w \in TM$;
  (iii) $\nabla(fX)(v) = f \nabla X(v) + df(v)X$, for all smooth $f : M \to \mathbf{R}$.

Exercise 13: *Extend the Levi-Civita connection ∇ to an operator, also denoted ∇,*

$$\nabla : \mathcal{T}^p_q \longrightarrow \mathcal{T}^p_q \otimes T^* M = \mathcal{T}^{p+1}_q,$$

by the following rules:
  (i) $\nabla dx^i = -\Gamma^i_{jk} dx^j \otimes dx^k$;
  (ii) for $\partial_{x^I} = \partial_{x^{i_1}} \otimes \ldots \otimes \partial_{x^{i_q}}$, $\nabla(\partial_{x^I}) = \sum_j \partial_{x^{i_1}} \otimes \ldots \otimes \nabla \partial_{x^{i_j}} \otimes \ldots \otimes \partial_{x^{i_q}}$, and similarly for $\nabla(dx^J)$;
  (iii) for any tensors ω, η, $\nabla(\omega \otimes \eta) = \nabla \omega \otimes \eta + \omega \otimes \nabla \eta$;
  (iv) for $f \in \mathcal{T}^0_0 = C^\infty(M)$, $\nabla f = df$.
Note that (iii), (iv) imply

$$\nabla(a^I_J \partial_{x^I} \otimes dx^J) = da^I_J \otimes \partial_{x^I} \otimes dx^J + a^I_J \nabla(\partial_{x^I} \otimes dx^J).$$

The important point is to check that these rules define an operator independent of the choice of local coordinates. Classically, these extensions of the Levi-Civita connection are written

$$\nabla(a^{i_1,\ldots,i_q}_{j_1,\ldots,j_p} \partial_{x^I} \otimes dx^J) = a^{i_1,\ldots,i_q}_{j_1,\ldots,j_p;j} \partial_{x^I} \otimes dx^J \otimes dx^j,$$

with corresponding covariant derivatives

$$\nabla_j(a_{j_1,\dots,j_p}^{i_1,\dots,i_q}\partial_{x^I}\otimes dx^J) = a_{j_1,\dots,j_p;j}^{i_1,\dots,i_q}\partial_{x^I}\otimes dx^J.$$

(The minus sign in (i) is forced by the equation

$$X(\omega(Z)) = (\nabla_X\omega)(Z) + \omega(\nabla_X Z)$$

for vector fields X, Z and a one-form ω. This reduces to the first equation in (2.24) if ω is identified with the vector field Y via the metric. Letting $\omega = dx^i, X = \partial_{x^j}, Z = \partial_{x^k}$ shows that we want a minus sign in (i).)

Now extend the Levi-Civita connection to k-forms, $\nabla : \Lambda^k T^*M \to T^*M \otimes \Lambda^k T^*M$, by imposing the Leibniz rule and

$$\nabla(\omega^1 \wedge \dots \wedge \omega^k) = \sum_i \omega^1 \wedge \dots \wedge \nabla\omega^i \wedge \dots \wedge \omega^k,$$

for one-forms ω^i.

Now consider the operator $\nabla^2 : C^\infty(M) \to T_0^2$. Recall from Chapter 1 that the metric g defines an isomorphism $\alpha = \alpha_{g,x} : T_xM \to T_x^*M$, and so a map $\mathrm{Tr} : T_x^*M \otimes T_x^*M \to T_x^*M \otimes T_xM \to \mathbf{R}$, where the last map is given by $v \otimes w \mapsto v(w)$. Check that $\Delta = -\mathrm{Tr}(\nabla^2)$ is the Laplacian on functions.

Since the Levi-Civita connection measures the change in vectors lying in different tangent spaces, it should provide a method of comparing (or "connecting") different tangent spaces. Indeed, given $x, y \in M$ and a curve $\gamma : [0, 1] \to M$ from x to y, we can define an isomorphism $||_\gamma : T_xM \to T_yM$ as follows. For $v \in T_xM$, consider the ODE for a vector field V defined in a neighborhood of γ:

$$\nabla_{\dot\gamma(t)}V(t) = 0, \ V(0) = v,$$

where $V(t)$ denotes $V(\gamma(t))$. Since ∇ is a first order differential operator, there is a unique solution $V(t)$. We set $||_\gamma(v) = V(1)$. The vector $V(1)$ is called the *parallel translation* of v along γ, as it is easy to check that in Euclidean space $V(1)$ is the same as v. Parallel translation is clearly an isomorphism, with inverse given by parallel translation along γ run in the reverse order. Note that parallel translation is independent of γ in Euclidean space, but this is not true for general Riemannian manifolds. Thus the Levi-Civita connection allows us to compare different tangent spaces, but only after a choice of γ.

Exercise 14: *(i) If $\gamma(t) = (\gamma^1(t),\dots,\gamma^n(t))$ and $V = v^k\partial_k$ in a coordinate chart, show that the equation $\nabla_{\dot\gamma(t)}V(t) = 0$ becomes the equations*

$$\frac{dv^k}{dt} + \frac{d\gamma^i}{dt}\Gamma_{ij}^k(\gamma(t))v^j(t) = 0, \quad k = 1,\dots,n,$$

for $v^k(t) = v^k(\gamma(t))$. In particular, the equation $\nabla_{\dot\gamma(t)}\dot\gamma(t) = 0, \gamma(0) = x_0, \dot\gamma(0) = v_0$ for fixed $x_0 \in M, v_0 \in T_{x_0}M$, is a second order differential equation (in a *coordinate neighborhood of x_0) and so has a unique solution for small t. (Strictly speaking, we must extend $\dot{\gamma}$ from a vector field just defined on γ to a vector field defined in a neighborhood of γ. However, if we let γ be the x^1 axis in a coordinate system, the expression for the covariant derivative in terms of the Christoffel symbols easily shows that $\nabla_{\dot{\gamma}(t)}\dot{\gamma}(t)$ is independent of this extension.) We will see in §2.3 that this equation is satisfied by geodesics on M.*

(ii) Show that parallel translation is an isometry. Thus parallel translation $||_\gamma$ around a closed loop can be considered to be an orthogonal transformation of T_xM. Hint: it is enough to show that $\dot{\gamma}\langle V(t), V(t)\rangle = 0$.

(iii) Define the holonomy group at $x \in M$ to be the subgroup of $O(n)$ given by parallel translations around all closed loops at x. Show that if M is simply connected, then the holonomy group actually lies in $SO(n)$. Show that for the standard sphere, the holonomy group at an arbitrary point is all of $SO(n)$.

Parallel translation can be extended to give an isomorphism $||_\gamma : \Lambda^k T_y^* M \to \Lambda^k T_x^* M$ as follows. For $k = 1$, set $||_\gamma$ to be the adjoint of $||_\gamma : T_xM \to T_yM$. For $k > 1$, define $||_\gamma$ to be the linear map extending

$$||_\gamma(v^1 \wedge \ldots \wedge v^k) = ||_\gamma v^1 \wedge \ldots \wedge ||_\gamma v^k,$$

for $v^1, \ldots, v^n \in T_y^* M$.

2.2.2 Weitzenböck Formulas and Gårding's Inequality

In this subsection we will derive the Weitzenböck formula relating the Laplacian on forms to the Levi-Civita connection. The Weitzenböck formula gives a proof of Bochner's theorem, relating $H^1_{dR}(M)$ to the Ricci curvature, and a proof of the basic elliptic estimate/Gårding's inequality of §1.3.4. These techniques will be used in Chapter 4 in the proof of the Chern-Gauss-Bonnet theorem.

We begin by showing that the Levi-Civita connection determines the curvature tensor. Since the curvature tensor R is a section of the bundle

$$TM \otimes T^*M \otimes T^*M \otimes T^*M \cong \mathrm{Hom}(TM \otimes TM \otimes TM, TM),$$

where this isomorphism does not use the metric, we can consider R as an operator taking three tangent vectors at a point to a fourth. Classically, this is denoted by $X, Y, Z \mapsto R(X,Y)Z$, for $X, Y, Z \in T_xM$. Note that $R(\partial_j, \partial_k)\partial_l = R^i_{jkl}\partial_i$.

Theorem 2.25 *Extend X, Y, Z to vector fields in a neighborhood of $x \in M$. Then*

$$R(X,Y)Z = \nabla_Y(\nabla_X Z) - \nabla_X(\nabla_Y Z) - \nabla_{[Y,X]}Z. \tag{2.26}$$

PROOF. If we replace X by fX, for any function f defined near x, it is easy to check that the right hand side of (2.26) is multiplied by f, and the same is true for Y, Z. Since $X = a^i\partial_i$ for functions a^i and similarly for Y, Z, we may assume $X = \partial_i, Y = \partial_j, Z = \partial_k$. The result follows by a direct computation.

We now introduce a linear algebra construction, called fermion calculus, a finite dimensional analogue of the creation and annihilation operators in elementary quantum mechanics. Let V be an n-dimensional inner product space. We won't distinguish between a vector in V and its dual in V^*. Associated to $v \in V$ are maps

$$a_v^* : \Lambda^p(V^*) \to \Lambda^{p+1}(V^*), \quad a_v : \Lambda^p(V^*) \to \Lambda^{p-1}(V^*),$$

given by $a_v^*(\omega) = v \wedge \omega$ (exterior multiplication) and by defining a_v to be interior product, the dual map to a_v^*.

Exercise 15: *Let $\{\theta_i\}$ be an orthonormal basis of V^*. Show that*

$$a_{\theta_i}(\theta_{j_1} \wedge \ldots \wedge \theta_{j_k}) = \begin{cases} 0, & i \notin \{j_1, \ldots, j_k\}, \\ (-1)^{q-1}\theta_{j_1} \wedge \ldots \wedge \hat{\theta}_{j_q} \wedge \ldots \wedge \theta_{j_k}, & i = j_q, \end{cases}$$

where $\hat{\theta}_{j_q}$ means that θ_{j_q} is omitted.

Let $\mathrm{End}(V^*)$ denote the space of endomorphisms of $\Lambda^*(V^*)$. For $A, B \in \mathrm{End}(V^*)$, define the anticommutator $\{A, B\}$ by $\{A, B\} = AB + BA$. For an orthonormal basis $\{\theta_i\}$ of V^*, let a_i, a_i^* denote $a_{\theta_i}, a_{\theta_i}^*$. With the help of the last exercise, it is easy to show that we have the basic relations

$$\{a_i, a_j\} = 0, \quad \{a_i^*, a_j^*\} = 0, \quad \{a_i, a_j^*\} = \delta_{ij}. \tag{2.27}$$

Consider the set of endomorphisms of the form

$$A_{IJ} = a_{i_1}^* \circ \ldots \circ a_{i_k}^* \circ a_{j_1} \circ \ldots \circ a_{j_\ell}, \tag{2.28}$$

with $1 \le i_1 < \ldots < i_k \le n$, $1 \le j_1 < \ldots < j_\ell \le n$ (with the understanding that the multi-indices I or J may be empty and that $A_{\emptyset\emptyset} = \mathrm{Id}$). There are 2^{2n} of these endomorphisms, which is the dimension of $\mathrm{End}(V^*)$, and we claim that in fact $\{A_{IJ}\}$ is a basis of $\mathrm{End}(V^*)$.

It is enough to show that $\{A_{IJ}\}$ is linearly independent. Say $c^{IJ}A_{IJ} = 0$. Since $A_{IJ} : \Lambda^p(V^*) \to \Lambda^{p+|I|-|J|}(V^*)$, we may assume that $|I| - |J|$ is constant. We now induct on the order of J. If $|J| = 0$, we obtain $0 = c^{IJ}A_{IJ}(1) = c^{I\emptyset}\theta_I$, where $\theta_I = \theta_{i_1} \wedge \ldots \wedge \theta_{i_k}$, and so $c^{I\emptyset} = 0$ for all I. By induction, we have $c^{IJ} = 0$ whenever $|J| < s$. Applying $c^{IJ}A_{IJ}$ to θ_{J_1}, where $|J_1| = s$, gives $c^{IJ_1} = 0$ for all I. Letting J_1 vary over all multi-indices of order s finishes the induction.

Lemma 2.29 *(i)* $\sum_i a_i^* a_i = p \cdot \mathrm{Id}$ *on* $\Lambda^p(V^*)$.
   (ii) Let $(-1)^F = (-1)^p \, \mathrm{Id} : \Lambda^p(V^*) \to \Lambda^p(V^*)$. *Then*

$$(-1)^F = \prod_i (1 - 2a_i^* a_i) = \prod_i (a_i a_i^* - a_i^* a_i).$$

PROOF. (i) It is easy to check that $a_i^* a_i(\theta_I) = \theta_I$ if $i \in I$ and zero otherwise. Since θ_I has p indices, $\sum_i a_i^* a_i(\theta_I) = p\theta_I$, and the result follows.

(ii) By (i), $(1 - 2a_i^* a_i)(\theta_I)$ equals θ_I if $i \notin I$ and $-\theta_I$ if $i \in I$. Thus

$$\prod_i (1 - 2a_i^* a_i)(\theta_I) = (-1)^p \theta_I,$$

and the last relation in (ii) follows from (2.27).

Exercise 16: *For $A \in \mathrm{End}(V)$, let $A^* : V^* \to V^*$ be the adjoint map, and define the induced endomorphism $A^* \in \mathrm{End}(\Lambda^* V^*)$ by*

$$A^*(\theta_I) = \sum_{j=1}^{k} \theta_{i_1} \wedge \ldots \wedge A^* \theta_{i_j} \wedge \ldots \wedge \theta_{i_k}.$$

Show that if $A = (A_{ij})$ is skew-symmetric, then A^* is given by $-A_{ij} a_i^* a_j$ (with summation over i, j).

For $A \in \mathrm{End}(V^*)$, we define the *supertrace* of A to be $\mathrm{Tr}((-1)^F A)$. This is just the trace of A on even forms minus the trace of A on odd forms. Such a trace occurs in the proof of the Chern-Gauss-Bonnet theorem in Chapter 4. Other common notation for the supertrace is $\mathrm{Tr}_s(A)$.

Proposition 2.30 *Let $A = c^{IJ} A_{IJ} \in \mathrm{End}(V^*)$. Then*

$$\mathrm{Tr}((-1)^F A) = (-1)^{n(n-1)/2} c^{\{1,2,\ldots,n\}\{1,2,\ldots,n\}}.$$

In other words, only the term in A with n a's and n a^*'s contributes to the supertrace of A.

PROOF. We introduce the "Clifford algebra variables/Dirac matrices"

$$e_i = a_i - a_i^*, \quad \bar{e}_i = a_i + a_i^*, \tag{2.31}$$

which satisfy the relations

$$\{e_i, e_j\} = -\{\bar{e}_i, \bar{e}_j\} = -2\delta_{ij}, \quad \{e_i, \bar{e}_j\} = 0. \tag{2.32}$$

We have

$$(-1)^F = \prod (a_i - a_i^*)(a_i + a_i^*) = (-1)^{n(n-1)/2} \gamma \bar{\gamma},$$

where $\gamma = e_1 e_2 \ldots e_n$ and $\bar{\gamma} = \bar{e}_1 \bar{e}_2 \ldots \bar{e}_n$. Now consider the endomorphism $B = \epsilon_{i_1} \epsilon_{i_2} \ldots \epsilon_{i_\ell}$ with ϵ_i equal to e_i or $\bar{e}_i$, and let $C = \epsilon_{i_2} \ldots \epsilon_{i_\ell}$. If ℓ is odd, then (2.32) implies $B\gamma\bar{\gamma} = -\gamma\bar{\gamma}B$, and so $\mathrm{Tr}(B\gamma\bar{\gamma}) = -\mathrm{Tr}(\gamma\bar{\gamma}B) = -\mathrm{Tr}(B\gamma\bar{\gamma})$ implies $\mathrm{Tr}(\gamma\bar{\gamma}B) = 0$. If ℓ is even, then $B = \epsilon_{i_1} C = -C\epsilon_{i_1}$, and so $\mathrm{Tr}(B) = 0$ unless $\ell = 0$. Thus $\mathrm{Tr}((-1)^F B) = 0$ unless $B = (-1)^{n(n-1)/2} \gamma\bar{\gamma}$. In this case,

$$\mathrm{Tr}((-1)^F (-1)^{n(n-1)/2} \gamma\bar{\gamma}) = \mathrm{Tr}(\mathrm{Id}) = 2^n, \tag{2.33}$$

since dim $V^* = 2^n$.

Since

$$a_i = \frac{1}{2}(e_i + \bar{e}_i), \quad a_i^* = -\frac{1}{2}(e_i - \bar{e}_i),$$

we can write any A_{IJ} as an expression in the $e_i, \bar{e}_i$. If I and J do not equal $\{1, \ldots, n\}$, then no term in this expression involves $\gamma\bar{\gamma}$, and so $\mathrm{Tr}((-1)^F A_{IJ}) = 0$. Finally, since $a_i a_i^* = (1/2)(1 + e_i \bar{e}_i)$,

$$
\begin{aligned}
A_{\{1,\ldots,n\}\{1,\ldots,n\}} &= (-1)^{n(n-1)/2} a_1^* a_1 a_2^* a_2 \ldots a_n^* a_n \\
&= (-1)^{n(n-1)/2} \prod \left(\frac{1}{2}(1 + e_i \bar{e}_i)\right) \\
&= (-1)^{n(n-1)/2} 2^{-n} \prod e_i \bar{e}_i + \ldots \\
&= (-1)^{n(n-1)/2} 2^{-n} (-1)^{n(n-1)/2} \gamma\bar{\gamma} + \ldots,
\end{aligned}
$$

where the dots indicate other products of the $e_i, \bar{e}_j$. Thus

$$\mathrm{Tr}((-1)^F A_{\{1,\ldots,n\}\{1,\ldots,n\}}) = 2^{-n} \mathrm{Tr}((-1)^F \gamma\bar{\gamma}) = (-1)^{n(n-1)/2}.$$

We now carry this linear algebra over to a Riemannian manifold M by replacing V with $T_x M$. On a neighborhood U of x, there exists a smoothly varying set of one-forms $\{\theta_i\}$ which form an orthonormal basis of $T_y^* M$, for all $y \in U$. (For example, just apply Gram-Schmidt to the basis $\{dx^i\}$ defined on a chart near x.) This orthonormal basis, or frame, defines endomorphisms $a_i, a_i^* \in \mathrm{End}(\Lambda^* T_y^* M)$ for $y \in U$. Define the *curvature endomorphism* by

$$R = -R_{ijkl} a_i^* a_j a_k^* a_l, \tag{2.34}$$

with summation convention. Here the components R_{ijkl} are computed with respect to the orthonormal frame. More precisely, as in (2.26) we have $R^i_{jkl} = R_{ijkl} = \langle R(X_i, X_k)X_\ell, X_i \rangle$, where $X_i \in T_y M$ is dual to θ_i, We will see below that the expression for R is independent of choice of frame.

Exercise 17: *Show that for an orthonormal frame* $\omega^1, \ldots, \omega^n \in T_x^* M$,

$$\langle R(\omega^1 \wedge \ldots \wedge \omega^k), \omega^1 \wedge \ldots \wedge \omega^k \rangle = \sum_{i=1}^{k} \sum_{j=k+1}^{n} K(Y_i, Y_j),$$

where the Y_i are dual tangent vectors to the ω^i, and $K(Y_i, Y_j)$ denotes the sectional curvature of the Y_i, Y_j plane. Hint: Assuming that R is independent of orthonormal frame, we may set $\omega^i = \theta^i$.

We now rewrite R in terms of the $e_i, \bar{e}_i$. This allows us to compute the supertrace of R^k, the composition of R with itself k times, a key technical step in the proof of the Chern-Gauss-Bonnet theorem.

Lemma 2.35 *Assume* $\dim M = n = 2k$ *is even. Let s be the scalar curvature of M.*

(a) *We have* $R = \dfrac{s}{4} + \dfrac{1}{8} R_{ijk\ell} e_i e_j \bar{e}_k \bar{e}_\ell$.

(b) *The supertrace* $\mathrm{Tr}((-1)^F R^k)$ *equals*

$$\frac{1}{2^k} \sum_{\sigma, \tau \in \Sigma_n} (\mathrm{sgn}\ \sigma)(\mathrm{sgn}\ \tau) R_{\sigma(1)\sigma(2)\tau(1)\tau(2)} \cdot \ldots \cdot R_{\sigma(n-1)\sigma(n)\tau(n-1)\tau(n)}.$$

Here Σ_n is the group of permutations of $\{1, \ldots, n\}$ and $\mathrm{sgn}\ \sigma$ denotes the sign of the permutation σ.

PROOF. (a) We claim that

$$R = -\frac{1}{4} R_{ijk\ell}(a_i^* a_j + a_i a_j^*)(a_k^* a_\ell + a_k a_\ell^*).$$

For example,

$$R_{ijk\ell} a_i^* a_j a_k a_\ell^* = -R_{ijk\ell} a_i^* a_j a_\ell^* a_k = -R_{ij\ell k} a_i^* a_j a_k^* a_\ell = R_{ijk\ell} a_i^* a_j a_k^* a_\ell,$$

as $R_{ijkk} = 0$ by (2.18). Similarly,

$$R_{ijk\ell} a_i a_j^* a_k^* a_\ell = R_{ijk\ell} a_i a_j^* a_k a_\ell^* = R_{ijk\ell} a_i^* a_j a_k^* a_\ell,$$

which gives the claim.

A direct computation using (2.31) gives

$$e_i e_j - \bar{e}_i \bar{e}_j = -2(a_i^* a_j + a_i a_j^*),$$

so

$$R = -\frac{1}{16} R_{ijk\ell}(e_i e_j - \bar{e}_i \bar{e}_j)(e_k e_\ell - \bar{e}_k \bar{e}_\ell).$$

If j, k, ℓ are distinct indices, the commutation relations (2.32) give $e_i e_j e_\ell = e_j e_\ell e_i = e_\ell e_i e_j$. Using the Bianchi identity (2.20), we see that for fixed i the sum over distinct j, k, ℓ of $R_{ijk\ell} e_i e_j e_k e_\ell$ vanishes.

In an orthonormal frame, the Ricci tensor is given by $R_{ik} = R_{ijkj}$ (as $g^{ij} = \delta^{ij}$). Moreover, by (2.18) $R_{ik} = R_{ki}$. Thus

$$
\begin{aligned}
R_{ijk\ell} e_i e_j e_k e_\ell &= R_{ijj\ell} e_i e_j e_j e_\ell + R_{ijkj} e_i e_j e_k e_j \\
&= -R_{ijj\ell} e_i e_\ell + R_{ijkj} e_i e_k \\
&= R_{ijkj} e_i e_k + R_{ijkj} e_i e_k = 2 R_{ik} e_i e_k \\
&= -2 R_{ii} + 2 \sum_{i \neq k} R_{ik} e_i e_k \\
&= -2s + 2 \sum_{i < k} R_{ik}(e_i e_k + e_k e_i) \\
&= -2s.
\end{aligned}
$$

Similarly, $R_{ijk\ell}\bar{e}_i\bar{e}_j\bar{e}_k\bar{e}_\ell = -2s$. This gives

$$R = -\frac{1}{16}(-4s - R_{ijk\ell}e_ie_j\bar{e}_k\bar{e}_\ell - R_{ijk\ell}\bar{e}_i\bar{e}_je_ke_\ell).$$

Again by (2.18), (2.32), we get

$$R_{ijk\ell}\bar{e}_i\bar{e}_je_ke_\ell = R_{ijk\ell}e_ke_\ell\bar{e}_i\bar{e}_j = R_{k\ell ij}e_ie_j\bar{e}_k\bar{e}_\ell = R_{ijk\ell}e_ie_j\bar{e}_k\bar{e}_\ell.$$

Thus

$$R = -\frac{1}{16}(-4s - 2R_{ijk\ell}e_ie_j\bar{e}_k\bar{e}_\ell).$$

(b) As in the proof of Proposition 2.30, especially (2.33), we must pick out the coefficient of $\gamma\bar{\gamma}$ in R^k. The only terms in R^k contributing to this coefficient are of the form

$$\frac{1}{8^k}R_{\sigma(1)\sigma(2)\tau(1)\tau(2)}\cdots R_{\sigma(n-1)\sigma(n)\tau(n-1)\tau(n)}e_{\sigma(1)}e_{\sigma(2)}\bar{e}_{\tau(1)}\bar{e}_{\tau(2)}\cdots\bar{e}_{\tau(n-1)}\bar{e}_{\tau(n)},$$

$$(2.36)$$

as any term with a repeated e_i or $\bar{e}_i$ will not contribute. Rearranging the product of the $e_{\sigma(i)}, \bar{e}_{\tau(j)}$ terms into the sequence

$$e_{\sigma(1)}\cdots e_{\sigma(n)}\bar{e}_{\tau(1)}\cdots\bar{e}_{\tau(n)} \qquad (2.37)$$

does not involve any change of sign, as each $\bar{e}_{\tau(j)}$ moves past an even number of $e_{\sigma(i)}$. However, rearranging (2.37) into $\gamma\bar{\gamma}$ introduces a factor of $(\mathrm{sgn}\ \sigma)(\mathrm{sgn}\ \tau)$ $\times(-1)^{n(n-1)/2}$. Using (2.33), (2.36) gives

$$\mathrm{Tr}((-1)^F R^k) = \frac{2^n}{8^k}\sum_{\sigma,\tau}(\mathrm{sgn}\ \sigma)(\mathrm{sgn}\ \tau)R_{\sigma(1)\sigma(2)\tau(1)\tau(2)}\cdot\cdots\cdot R_{\sigma(n-1)\sigma(n)\tau(n-1)\tau(n)}.$$

As in Exercise 13, the Levi-Civita connection may be considered to be a map $\nabla : \Gamma(T^*M) \longrightarrow \Gamma(T^*M \otimes T^*M)$. Both the range and domain of ∇ have the global L^2 inner products of §1.2.2, once $T_x^*M \otimes T_x^*M$ is given the usual inner product $\langle v \otimes w, s \otimes t\rangle = \langle v, s\rangle\langle w, t\rangle$. Thus ∇ has a (formal) adjoint $\nabla^* : \Gamma(T^*M \otimes T^*M) \to \Gamma(T^*M)$, and we can form the second order differential operator $\nabla^*\nabla : \Lambda^1 T^*M \to \Lambda^1 T^*M$.

Exercise 18: *Show that the second order part of $\nabla^*\nabla$ equals $-g^{ij}\nabla_i\nabla_j$. Here $\nabla_i = \nabla_{X_i}$.*

It is an unpleasant exercise to compare Δ^1 and $\nabla^*\nabla$ in local coordinates:

Exercise 19: *Show that*

(i) $\Delta^1(a_j dx^j) = g^{ik}\left(\dfrac{\partial^2 a_j}{\partial x^i \partial x^k}\right)dx^j$ + (lower order terms);

(ii) $\nabla^*\nabla(a_j dx^j) = g^{ik}\left(\dfrac{\partial^2 a_j}{\partial x^i \partial x^k}\right)dx^j$ + (lower order terms).

Thus Δ^1 and $\nabla^*\nabla$ differ by a first order operator. In fact, the first order term is zero. We will show this in more generality. As in Exercise 13, we may consider ∇ as an operator on $\nabla : \Lambda^k T^*M \to \Lambda^k T^*M \otimes T^*M$ by the derivation formula

$$\nabla(\omega^1 \wedge \ldots \wedge \omega^k) = \sum_i \omega^1 \wedge \ldots \wedge \nabla\omega^i \wedge \ldots \wedge \omega^k,$$

for one-forms ω^i. Imposing the Leibniz rule then extends ∇ to all k-forms. Taking a direct sum over k gives an operator ∇ on $\Lambda^* T^*M$, the space of all forms of mixed degree. As above, ∇ has a formal adjoint ∇^*, and we call $\nabla^*\nabla$ the Bochner Laplacian.

Theorem 2.38 (Weitzenböck formula) *The Laplacian on forms satisfies*

$$\Delta = \nabla^*\nabla + R.$$

To prove the theorem, it is convenient to work in a synchronous frame. This is given by first filling out a neighborhood of x with lines radiating from x and intersecting only at x. (These lines are usually chosen to be geodesics, as defined in the next section, but that is not necessary here; the lines can be chosen to be the image of the radial lines of a polar coordinate system in $\mathbf{R}^n$ under the coordinate map.) Next, an orthonormal frame $\{X_i\}$ of $T_x M$ is chosen and then parallel translated out the radial lines. Since $0 = \nabla_{X_i}X_j = \Gamma_{ij}^k X_k$ at x, we have $\Gamma_{ij}^k(x) = 0$ for all i, j, k.

Lemma 2.39 *Let ∇_i denote ∇_{X_i}, with X_i as above. Then we have*

$$d = \sum_i a_i^* \nabla_i, \quad \delta = d^* = \sum_i \nabla_i^* a_i, \quad \nabla^*\nabla = \sum_i \nabla_i^* \nabla_i,$$

and at the center point x of a synchronous frame, we have

$$\nabla_i^* = -\nabla_i, \quad \Delta = -\sum_{k,\ell}(a_k^* a_\ell \nabla_k \nabla_\ell + a_\ell a_k^* \nabla_\ell \nabla_k).$$

PROOF OF THE LEMMA. Let $\{\theta^i\}$ be the orthonormal frame of one-forms dual to $\{X_i\}$. Using [64, I.7-13] to compute the differential of a one-form, we get

$$\begin{aligned} d\theta^i(X_j, X_k) &= X_j(\theta^i(X_k)) - X_k(\theta^i(X_j)) - \theta^i([X_j, X_k]) \\ &= 0 - 0 - \theta^i(\nabla_{X_j}X_k - \nabla_{X_k}X_j) \\ &= \Gamma_{jk}^i - \Gamma_{kj}^i, \end{aligned}$$

and so $d\theta^i = \Gamma_{jk}^i \theta^j \wedge \theta^k$. Thus for $|I| = r$,

$$d\theta^I = \sum (-1)^{s-1}\theta^{i_1} \wedge \ldots \wedge d\theta^{i_s} \wedge \ldots \wedge \theta^{i_r} = \Gamma_{jk}^{i_s} a_j^* a_k^* a_{i_s} \theta^I.$$

Since $d = \nabla$ on functions, we have $d\alpha_I = \nabla(\alpha_I) = \nabla_i(\alpha_I)\theta^i = a_i^*\nabla_i(\alpha_I)$. This implies

$$
\begin{aligned}
d(\alpha_I\theta^I) &= d\alpha_I \wedge \theta^I + \alpha_I d\theta^I \\
&= a_i^*\nabla_i(\alpha_I)\theta^I + \alpha_I\Gamma_{jk}^i a_j^* a_k^* a_{i_*}\theta^I \\
&= a_i^*\nabla_i(\alpha_I\theta^I) - \alpha_I a_i^*\nabla_i\theta^I + \alpha_I\Gamma_{jk}^i a_j^* a_k^* a_{i_*}\theta^I \\
&= a_i^*\nabla_i(\alpha_I\theta^I) - \alpha_I a_i^*\Gamma_{jk}^i a_j^* a_k^* a_{i_*}\theta^I + \alpha_I a_i^*\Gamma_{jk}^i a_j^* a_k^* a_{i_*}\theta^I \\
&= a_i^*\nabla_i(\alpha_I\theta^I),
\end{aligned}
$$

which proves the first statement. The second statement follows immediately.

For $\nabla^*\nabla$, we note that

$$
\begin{aligned}
\langle\nabla^*\nabla\omega,\eta\rangle &= \langle\nabla\omega,\nabla\eta\rangle = \langle\nabla_i\omega\otimes\theta^i,\nabla_j\eta\otimes\theta^j\rangle \\
&= \langle\nabla_i\omega,\nabla_i\eta\rangle = \langle\nabla_i^*\nabla_i\omega,\eta\rangle.
\end{aligned}
$$

Thus $\nabla^*\nabla = \nabla_i^*\nabla_i$.

We now show that $\nabla_i^* = -\nabla_i$ at x. For $\omega = \alpha_I\theta^I$, we have

$$
\langle\omega,\nabla_i^*\eta\rangle = \langle\nabla_i\omega,\eta\rangle = \langle X_i(\alpha_I)\theta^I + \alpha_I\nabla_i\theta^I,\eta\rangle. \tag{2.40}
$$

As in Exercise 16, we have $\nabla_i\theta^I = \Gamma_{ik}^\ell a_k^* a_\ell\theta^I$, so the last term in (2.40) becomes

$$
\langle\alpha_I\Gamma_{ik}^\ell a_k^* a_\ell\theta^I,\eta\rangle = \langle\omega,\Gamma_{ik}^\ell a_\ell^* a_k\eta\rangle. \tag{2.41}
$$

For the first term on the right hand side of (2.40), we claim that

$$
\langle X_i(\alpha_I)\theta^I,\eta_J\theta^J\rangle = \langle\alpha_I\theta^I,(-X_i + \operatorname{div} X_i)(\eta_J)\theta^J\rangle.
$$

Assuming the claim, we see that (2.40) becomes

$$
\langle\omega,\nabla_i^*\eta\rangle = \langle\omega,(-X_i + \operatorname{div} X_i)(\eta_J)\theta^J + \Gamma_{ik}^\ell a_\ell^* a_k\eta_J\theta^J\rangle
$$

for $\eta = \eta_J\theta^J$, and so

$$
\nabla_i^*\eta = (-X_i + \operatorname{div} X_i)(\eta_J)\theta^J + \Gamma_{ik}^\ell a_\ell^* a_k\eta_J\theta^J. \tag{2.42}
$$

At x, $\Gamma_{ik}^\ell = 0$ and

$$
\operatorname{div} X_i = \delta\theta^i = -*d*\theta^i = \pm*d(\theta^1\wedge\ldots\wedge\theta^{i-1}\wedge\theta^{i+1}\wedge\ldots\wedge\theta^n) = 0
$$

as above. Thus $\nabla_i^*\eta = -(X_i\eta_J)\theta^J = -\nabla_i\eta$, where in the last step we use $\nabla_i\theta^I = 0$ at x.

For the claim, we write $X_i = a_i^j\partial_j = a_i^j\partial_{x^j}$ for local coordinates $\{x^j\}$ and compute

$$
\langle X_i(\alpha_I)\theta^I,\eta_J\theta^J\rangle = \int_M X_i(\alpha_I)\theta^I\wedge*\eta_J\theta^J
$$

$$= \int_M a_i^j \partial_j(\alpha_I)\eta_I \ \mathrm{dvol}$$

$$= -\int_M \alpha_I \partial_j(\eta_I a_i^j \sqrt{\det g})dx^1 \ldots dx^n$$

$$= -\int_M \alpha_I a_i^j \partial_j(\eta_I)\mathrm{dvol} - \int_M \alpha_I \eta_I \partial_j(a_i^j \sqrt{\det g})(\sqrt{\det g})^{-1} \ \mathrm{dvol}$$

$$= \int_M -\alpha_I X_i(\eta_I) + (\mathrm{div}\ X_i)(\eta_I) \ \mathrm{dvol}$$

$$= \int_M \alpha_I \theta^I \wedge *(-X_i + \mathrm{div}\ X_i)(\eta_J)\theta^J$$

$$= \langle \alpha_I \theta^I, (-X_i + \mathrm{div}\ X_i)(\eta_J)\theta^J,$$

where we abuse notation by ignoring the partition of unity needed in the integral.

For the last statement, we have $\Delta = d\delta + \delta d = a_j^* \nabla_j \nabla_i^* a_i + \nabla_i^* a_i a_j^* \nabla_j$. At x, we have $a_j^* \nabla_j = a_j^* \nabla_j$, $a_i \nabla_j = \nabla_j a_i$, etc., and so

$$\Delta = a_j^* a_i \nabla_j \nabla_i^* + a_i a_j^* \nabla_i^* \nabla_j. \tag{2.43}$$

Since $\nabla_i^* = -\nabla_i$ at x, we have $a_i a_j^* \nabla_i^* \nabla_j = -a_i a_j^* \nabla_i \nabla_j$ in (2.43), and the proof will be finished by showing $\nabla_j \nabla_i^* = -\nabla_j \nabla_i$ at x. First note that, as above, we have

$$\nabla_i(\eta_J \theta^J) = X_i(\eta_J)\theta^J + \eta_J \nabla_i \theta^J = X_i(\eta_J)\theta^J + \Gamma_{ik}^\ell a_\ell a_k^* \eta_J \theta^J.$$

By this equation and (2.42), we have

$$\nabla_j \nabla_i^* \eta = -\nabla_j(X_i(\eta_J)\theta^J) + \nabla_j(\mathrm{div}\ X_i + \Gamma_{ik}^\ell a_\ell a_k^*)\eta$$
$$= -\nabla_j(X_i(\eta_J)\theta^J) + (X_j \mathrm{div}\ X_i)\eta + (\mathrm{div}\ X_i)\nabla_j \eta + \nabla_j(\Gamma_{ik}^\ell a_\ell a_k^* \eta),$$
$$-\nabla_j \nabla_i \eta = -\nabla_j(X_i(\eta_J)\theta^J) - \nabla_j(\Gamma_{ik}^\ell a_\ell^* a_k \eta).$$

Since $\Gamma_{ik}^\ell = 0$ at x, it suffices to show that

$$(X_j \Gamma_{ik}^\ell)a_\ell a_k^* + X_j \mathrm{div}\ X_i + (\mathrm{div}\ X_i)\nabla_j \eta = -(X_j \Gamma_{ik}^\ell)a_k^* a_\ell.$$

From $\{a_k^*, a_\ell\} = \delta_{k\ell}$, this equation follows if we show $X_j \mathrm{div}\ X_i = -X_j \Gamma_{i\ell}^\ell$ and $\mathrm{div}\ X_i = 0$ at x. Now

$$\mathrm{div}\ X_i = \delta\theta^i = -*d*\theta^i = \pm *d(\theta^1 \wedge \ldots \wedge \theta^{i-1} \wedge \theta^{i+1} \wedge \ldots \wedge \theta^n)$$
$$= -*\Gamma_{i\ell}^\ell \ \mathrm{dvol},$$

where we have used $d\theta^i = \Gamma_{jk}^i \theta^j \wedge \theta^k$. Since $* \mathrm{dvol} = 1$, the proof is finished.

PROOF OF THE THEOREM. At the base point x of a synchronous frame, we have

$$\Delta = d\delta + \delta d = -\sum_{k,\ell}(a_k^* a_\ell \nabla_k \nabla_\ell + a_\ell a_k^* \nabla_\ell \nabla_k)$$
$$= -\sum_{k,\ell}\{a_k^*, a_\ell\}\nabla_k \nabla_\ell - a_\ell a_k^*(\nabla_k \nabla_\ell - \nabla_\ell \nabla_k).$$

Recall that for the Levi-Civita connection, we have $\nabla_X Y - \nabla_Y X = [X, Y]$, and so at x we get $[X_i, X_j] = 0$. By Theorem 2.25,

$$(\nabla_k \nabla_\ell - \nabla_\ell \nabla_k)(X) = R(X_k, X_\ell)X$$

for any $X \in \Lambda^p T_x^* M$, where the skew-symmetric endomorphism $R(X_k, X_\ell)$ of $T_x M$ is extended to an endomorphism of $\Lambda^p T_x^* M$ as in Exercise 16, namely $R(X_k, X_\ell) = -\sum_{i,j} R_{ijk\ell} a_i^* a_j$. By (2.27), we get

$$\Delta = -\sum_k \nabla_k \nabla_k + R = \nabla^* \nabla + R.$$

Since Δ and $\nabla^* \nabla$ are independent of frame, so is R.

The reader may wonder why we chose to work in a synchronous frame, since differential geometry is supposed to be a coordinate free subject (and the frame is not even a basis of $T_x M$ associated to any coordinates). The point is that while tensors are coordinate free quantities, the Christoffel symbols and the ∂_i are not tensors (although their combination into ∇_i is), and so some choice of coordinate chart or frame may simplify calculations involving $\Gamma_{jk}^i, \partial_i$. The unconvinced reader should check the derivation of the Weitzenböck formula in general coordinates in [19].

The Weitzenböck formula is particularly interesting for the case of one-forms, as then the relation $a_j a_k^* = \delta_{jk} - a_k^* a_j$ implies

$$-R_{ijk\ell} a_i^* a_j a_k^* a_\ell = R_{i\ell} a_i^* a_\ell - R_{ijk\ell} a_i^* a_k^* a_j a_\ell,$$

where $R_{i\ell}$ are the components of the Ricci tensor computed in the frame $\{X_i\}$. Since $a_j a_\ell = 0$ on one-forms, only the Ricci term remains. Moreover, for each $x \in M$ we can consider the Ricci curvature Ric_x as an element of $\text{Hom}(T_x^* M, T_x^* M)$ via the isomorphisms

$$\begin{aligned}
\text{Ric}_x &= R_{ij} \theta^i \otimes \theta^j \in \text{Hom}(T_x M \otimes T_x M, \mathbf{R}) \cong \text{Hom}(T_x M, T_x^* M \otimes \mathbf{R}) \\
&\cong \text{Hom}(T_x M, T_x^* M) \cong \text{Hom}(T_x^* M, T_x^* M),
\end{aligned}$$

where the last isomorphism uses the metric. Under these isomorphisms, Ric_x corresponds to $R_{i\ell} a_i^* a_\ell$, so we will just denote the last expression by Ric. This yields

Theorem 2.44 (Bochner's Formula) $\Delta^1 = \nabla^* \nabla + \text{Ric}$.

This gives our first nontrivial application of Hodge theory.

Theorem 2.45 (Bochner's Theorem) *If M is compact and oriented, and if* $\text{Ric} > 0$, *then* $H_{dR}^1(M) = 0$.

PROOF. If $H_{dR}^1(M) \neq 0$, then by the Hodge theorem there exists a nonzero $\omega \in \text{Ker } \Delta^1$. We have

$$0 = \langle \Delta\omega, \omega \rangle = \langle \nabla^*\nabla, \omega \rangle + \langle \text{Ric } \omega, \omega \rangle = \langle \nabla\omega, \nabla\omega \rangle + \int_M \langle \text{Ric } \omega, \omega \rangle \, \text{dvol}(x).$$

The first term on the right hand side is nonnegative, while the integrand is positive on some open subset of M where ω is nonzero. Thus the right hand side of the equation is positive, a contradiction.

Thus a manifold M with $H_{dR}^1(M) \neq 0$ admits no metric with positive Ricci curvature.

Exercise 20: *(i) Use Bochner's theorem to re-prove parts of Corollary 2.7. Namely, show that neither T^2 nor an oriented surface of genus $g > 1$ admits a metric of positive Gaussian curvature. Note that in fact we can replace "positive Gaussian curvature" by "nonnegative Gaussian curvature which is positive at some point."*

(ii) Complete the argument re-proving Corollary 2.7(iii) as follows. If the surface Σ of genus $g > 1$ admits a metric with vanishing Gaussian curvature, then by Bochner's theorem every harmonic one-form ω satisfies $\nabla\omega = 0$. This is a first order differential equation for ω and thus ω is determined by its "initial value" at any fixed point of Σ. Thus $\dim H_{dR}^1(\Sigma) \leq 2$. However, it is well known that $\chi(\Sigma) = 2 - 2g$, and so $\dim H_{dR}^1(\Sigma) = 2g$.

(iii) Use the argument in (ii) to show that a manifold M^n with nonnegative Ricci curvature has $\dim H_{dR}^1(M) \leq n$, and that a flat manifold has $\dim H_{dR}^k(M) \leq \binom{n}{k}$. Show that this result is sharp by considering the n-torus with the flat metric induced from $\mathbf{R}^n$.

(iv) We'll see in Chapter 4 that flat manifolds have vanishing Euler characteristic. This is not very interesting in odd dimensions. However, show that $\Sigma \times S^1$ admits no metric with nonnegative Ricci curvature, where Σ is a surface with Euler characteristic less than zero. In particular, this manifold admits no flat metric.

Bochner's theorem, which dates from around 1945, is actually weaker than Myers' theorem, from around 1940. Indeed, it follows from the discussion after Exercise 38, Chapter 1, that a manifold with finite fundamental group has $0 = H_{sing}^1(M; \mathbf{R}) \cong H_{dR}^1(M)$. (The converse is not true.)

As promised in §1.3.4, the Weitzenböck formula leads to a relatively simple proof of Gårding's inequality for the Laplacian on forms.

Theorem 2.46 *For all $s \in \mathbf{Z}^+ \cup \{0\}$, there exists a positive constant C_s such that*

$$\|\omega\|_{s+1} \leq C_s(\|\omega\|_s + \|(d+\delta)\omega\|_s)$$

for all $\omega \in H_{s+1}\Lambda^k$.

PROOF. (cf. [59, p. 42]) Let $\omega \in \Lambda^* T^* M$ denote a form of mixed degree. By a partition of unity argument, we may assume that ω has support contained in a coordinate chart. We will do induction on s and begin with $s = 0$.

Since $\Delta = (d + \delta)(d + \delta) = (d + \delta)^*(d + \delta)$, we have

$$
\begin{aligned}
\|(d + \delta)\omega\|_0^2 &= \langle (d + \delta)\omega, (d + \delta)\omega \rangle = \langle \Delta\omega, \omega \rangle \\
&= \langle (\nabla^*\nabla + R)\omega, \omega \rangle = \|\nabla\omega\|_0^2 + \langle R\omega, \omega \rangle.
\end{aligned}
$$

Since R is bounded below on M, there is a constant C_1 such that

$$
\|\nabla\omega\|_0^2 \leq C_1(\|\omega\|_0^2 + \|(d + \delta)\omega\|_0^2). \tag{2.47}
$$

Now let us compare $\|\nabla\omega\|_0^2$ with $\|\omega\|_1^2$. For $\omega = \omega_I dx^I$, we have

$$
\begin{aligned}
\|\nabla\omega\|_0^2 &= \int_M \langle \nabla\omega, \nabla\omega \rangle \, \text{dvol} \\
&= \int_M \langle \nabla_i\omega \otimes dx^i, \nabla_j\omega \otimes dx^j \rangle \, \text{dvol} = \int_M g^{ij} \langle \nabla_i\omega, \nabla_j\omega \rangle \, \text{dvol}.
\end{aligned}
$$

Writing $\nabla_i = \partial_i - \Gamma_i$ as in the proof of Lemma 2.39, we get

$$
\begin{aligned}
\|\nabla\omega\|_0^2 &= \int_M g^{ij} \langle (\partial_i - \Gamma_i)\omega, (\partial_j - \Gamma_j)\omega \rangle \, \text{dvol} \\
&= \int_M g \cdot \partial_i\omega_I \partial_j\omega_J \, \text{dvol} - \int_M g \cdot \Gamma_i\omega_I(2\partial_j - \Gamma_j)\omega_J \, \text{dvol}, \tag{2.48}
\end{aligned}
$$

where g denotes some products of $g^{k\ell}$ giving rise to the positive definite inner products $\langle \,,\, \rangle_x$ on $\Lambda^* T_x^* M$. Estimating the last term in (2.48) by Cauchy-Schwarz gives

$$
\|\nabla\omega\|_0^2 \geq (\|\omega\|_1^2 - \|\omega\|_0^2) - C_2\|\omega\|_0\|\omega\|_1, \tag{2.49}
$$

for some constant C_2, since the coefficients Γ_j are bounded above on M. Now given $\epsilon > 0$, there exists $K > 0$ such that for all $a, b > 0$, we have $ab \leq \epsilon a^2 + Kb^2$. Thus we can write $C_2\|\omega\|_0\|\omega\|_1 \leq (1/2)\|\omega\|_1^2 + C_3\|\omega\|_0^2$. Plugging this in (2.49) gives

$$
\|\nabla\omega\|_0^2 \geq \frac{1}{2}\|\omega\|_1^2 - C_3\|\omega\|_0^2. \tag{2.50}
$$

Combining (2.47), (2.50) gives a constant C_4 with

$$
\begin{aligned}
\|\omega\|_1^2 &\leq C_4(\|\omega\|_0^2 + \|(d + \delta)\omega\|_0^2) \\
&\leq C_4(\|\omega\|_0^2 + 2\|\omega\|_0\|(d + \delta)\omega\|_0 + \|(d + \delta)\omega\|_0^2),
\end{aligned}
$$

and taking the square root of this inequality finishes the proof for $s = 0$.

Fortunately, the induction step is easy. Denote $d + \delta$ by D. We have

$$
\|\omega\|_{s+1} \leq \sum_i \|\partial_i\omega\|_s \leq C_{s-1}(\|\partial_i\omega\|_{s-1} + \|D\partial_i\omega\|_{s-1})
$$

by induction. Since ∂_i is a first order differential operator, $\|\partial_i\omega\|_{s-1} \leq A_1\|\omega\|_s$. Moreover, $[D, \partial_i] = D\partial_i - \partial_i D$ is apparently a second order operator, but it is easy to check that the highest order differentiations cancel, and so $[D, \partial_i]$ is only first order. Thus by the triangle inequality,

$$\begin{aligned} \|D\partial_i\omega\|_{s-1} &\leq \|\partial_i D\omega\|_{s-1} + \|[D, \partial_i]\omega\|_{s-1} \\ &\leq A_1\|D\omega\|_s + A_2\|\omega\|_s. \end{aligned}$$

We finally obtain

$$\|\omega\|_{s+1} \leq C_s(\|\omega\|_s + \|D\omega\|_s).$$

2.3 Geodesics

The second major topic Gauss addressed in his study of surfaces was the existence of curves of shortest length between two fixed points on the surface. It is important to note that finding a path of shortest length is a harder problem than the mapping problem for several reasons. First, the curve problem may have no solution: as mentioned before, if the surface is $\mathbf{R}^2 - \{(0,0)\}$ and the points in question are (x, y) and $(-x, -y)$, then there is no curve of shortest length between these points, as the best we can do is to construct curves of length arbitrarily close to $2\sqrt{x^2 + y^2}$. It turns out that this type of problem does not occur if the manifold is compact; shortest paths always exist.

Moreover, the mapping problem was a local problem, which we could expect to solve only in a neighborhood of a point. In contrast, finding the shortest path is a global problem, one which involves selecting nice paths from the huge space of all paths on the manifold which are fixed at the endpoints. Finally, the solution of the mapping problem reduces to standard PDE techniques. However, finding the shortest path on a Riemannian manifold will have to involve more than just calculus.

To understand this last point, consider two almost antipodal points on the standard S^2. There is a unique path of shortest length, a piece of a great circle, between them. Now consider a mountain slowly growing somewhere on this great circle; more precisely, consider a smooth perturbation of the embedding of the sphere in $\mathbf{R}^3$ with the corresponding family of induced metrics. As the metric on the now bumpy sphere changes, the shortest path between the points also changes. Moreover, depending on how the mountain grows, the shortest path between the two points may at some instant suddenly jump to a qualitatively new path, say the portion of the great circle on the original sphere going "around the back." In other words, the shortest path between two points may depend in a discontinuous way upon the parameter describing the family of metrics, and so calculus techniques alone will not be sufficient.

Although this sounds like a pathological situation, it in fact occurs often in ordinary calculus problems; the student is usually just not told about it. Consider the problem of finding the minima of each of the smooth functions

$f_t : \mathbf{R} \to \mathbf{R}$, for $t \in [0, 1]$. For fixed t, of course, we first find the candidates for the minima by solving $f_t'(x) = 0$. The art of writing calculus texts consists in picking functions such that this equation is solvable, and such that one can easily check the roots to determine the absolute minima; note that checking the second derivative only finds the local minima. The dependence of the functions on t causes new complications in that the locus of the minima may fail to be continuous in t:

Exercise 21: *Find a family of functions* $f_t(x), t \in [0, 1]$, *smooth in t and x such that there is no continuous function* $g : [0, 1] \to \mathbf{R}$ *such that* $g(t) \in \{x \in \mathbf{R} : f_t$ *has a global minimum at* $x\}$. *(Note that the noncompactness of* $\mathbf{R}$ *is not relevant here.)*

In summary, we have all been misled into believing that the study of minima of functions was purely a calculus problem. In studying paths of shortest length, these difficulties cannot be overlooked.

Nevertheless, as for ordinary functions the first step is applying calculus techniques in order to find the possible shortest paths between points x and y on a Riemannian manifold M. In our case, the function we want to minimize is the length function

$$l(\gamma) = \int_0^1 \langle \dot{\gamma}(t), \dot{\gamma}(t) \rangle^{\frac{1}{2}} dt,$$

for all paths $\gamma : [0, 1] \to M$ with $\gamma(0) = x, \gamma(1) = y$. Within a fixed coordinate chart, this is a standard calculus of variations problem, with the Euler-Lagrange equation, the equation for a critical path for the length function, given by a second order quasi-linear ODE in the parameter t. We will instead give a coordinate free derivation of the Euler-Lagrange equation. We call a path γ *critical* if for any family of paths $\gamma_\epsilon(t)$, with $\epsilon \in (-1/2, 1/2)$, $t \in [0, 1]$, $\gamma_\epsilon(0) = x, \gamma_\epsilon(1) = y$ and with $\gamma_0(t) = \gamma(t)$, we have $\left. \dfrac{d}{d\epsilon} \right|_{\epsilon=0} \ell(\gamma_\epsilon) = 0$. Note that every path of minimal length between x and y is critical.

Theorem 2.51 *A path* $\gamma(t)$ *is critical iff*

$$\nabla_{\dot{\gamma}(t)} \dot{\gamma}(t) = 0.$$

PROOF.   Define $F : [0, 1] \times (-1/2, 1/2) \to M$ by $F(t, \epsilon) = \gamma_\epsilon(t)$ and set $N(\gamma(t)) = dF(\partial_s)$, where ∂_s is the positively pointing unit tangent vector field to $(-1/2, 1/2)$. Note that N is a vector field defined along the image of γ if F is injective. In general, N assigns to a point (t, ϵ) a vector in $T_{F(t,\epsilon)}M$. Classically, N was called a parametrized vector field on M; in modern language, N is a section of F^*TM, the pullback of the tangent bundle TM by the map F. N is called the variation vector field of F.

Assume for simplicity that γ is parametrized by arclength and is of length one; in general, we just replace $\int_0^1$ in $\ell(\gamma)$ with $\int_0^{\ell(\gamma)}$. Then

$$
\begin{aligned}
0 &= \left.\frac{d}{d\epsilon}\right|_{\epsilon=0} \ell(\gamma_\epsilon(t)) = \int_0^1 \left.\frac{d}{d\epsilon}\right|_{\epsilon=0} \langle \dot\gamma_\epsilon(t), \dot\gamma_\epsilon(t)\rangle_{\gamma_\epsilon(t)}^{1/2}\, dt \\
&= \frac{1}{2}\int_0^1 N\left(\langle \dot\gamma_\epsilon(t), \dot\gamma_\epsilon(t)\rangle_{\gamma_\epsilon(t)}\right)|_{\epsilon=0}\, dt \\
&= \int_0^1 \langle \nabla_N \dot\gamma, \dot\gamma\rangle\, dt,
\end{aligned}
$$

by (2.24).

We have $0 = dF([\partial_t, \partial_s]) = [dF\partial_t, dF\partial_s] = [\dot\gamma, N]$. As above, this must be interpreted appropriately if F is not injective. Now (2.24) and the equation above give

$$
\begin{aligned}
0 &= -\int_0^1 \langle \nabla_{\dot\gamma} N, \dot\gamma\rangle\, dt \\
&= \int_0^1 \langle N, \nabla_{\dot\gamma}\dot\gamma\rangle\, dt - \int_0^1 \dot\gamma\langle N, \dot\gamma\rangle\, dt.
\end{aligned}
$$

The last integral is just

$$
-\int_0^1 \frac{d}{dt}\langle N_{(t,0)}, \dot\gamma(t)\rangle_{\gamma(t)}\, dt = \langle N, \dot\gamma\rangle\Big|_0^1 = 0,
$$

since $F(0,\epsilon) = x, F(1,\epsilon) = y$ for all ϵ implies $N_{(0,0)} = N_{(1,0)} = 0$.

It is easy to see that given a vector field N along γ, there is a variation F of γ with variation vector field N; the proof reduces to a linear integration in $\mathbf{R}^n$. Thus a path is critical iff for every vector field N along γ, we have

$$
0 = \int_0^1 \langle N, \nabla_{\dot\gamma}\dot\gamma\rangle\, dt.
$$

Letting $N = \nabla_{\dot\gamma}\dot\gamma$ shows that we must have $0 = \nabla_{\dot\gamma}\dot\gamma$ pointwise.

Definition: *A critical point for the length function is called a* geodesic.

Thus a geodesic is a curve with parallel tangent vectors, and as it must generalizes the case of straight lines in Euclidean space. The terminology comes from geodesy, the science of measurement of the earth. It is interesting to compare this derivation of the geodesic equation with the classical derivation in local coordinates (see [64, Vol. I]). One ends up with the Euler-Lagrange equation for $\gamma(t) = (\gamma^1(t), \ldots, \gamma^n(t))$ given by

$$
\frac{d^2\gamma^i}{dt^2} + \Gamma^i_{jk}\frac{d\gamma^j}{dt}\frac{d\gamma^k}{dt} = 0, \quad i = 1, \ldots n, \tag{2.52}
$$

in agreement with Exercise 14. Since a neighborhood of x is diffeomorphic to $\mathbf{R}^n$, the existence and uniqueness theory of ODEs guarantees that, for any $x \in M$ and $v \in T_x M$, there exist an $\epsilon > 0$ and a unique geodesic $\gamma_v(t)$, $t \in (-\epsilon, \epsilon)$, with $\gamma(0) = x, \gamma'(0) = v$.

Exercise 22: *Show that if M is compact (and without boundary), then $\gamma_v(t)$ exists for all $t \in \mathbf{R}$. Hint: Let $T_0 = \sup\{T \in \mathbf{R} : \gamma_v(t) \text{ exists on } (-T, T)\}$. If $T_0 < \infty$, show that $\lim_{t \to T_0} \gamma(t) = x_0$ and $\lim_{t \to T_0} \dot{\gamma}(t) = V \in T_{x_0} M$ exist. Then show that the geodesic $\gamma_V(t)$ starting at x_0 extends γ_v past T_0, giving a contradiction.*

Exercise 23: *(Hopf-Rinow theorem for compact manifolds; cf. [31, Ch. 1], [46, §10]) Recall from Exercise 9, Chapter 1, that $d(x, y) = \inf\{\ell(\gamma) : \gamma : [0, 1] \to M, \gamma(0) = x, \gamma(1) = y\}$ defines a metric on M. Let M be compact, and choose $x, y \in M$. Show that there exists a smooth curve γ from x to y with $\ell(\gamma) = d(x, y)$. Conclude that γ is a geodesic. Hint: Take a sequence of curves γ_n from x to y with $\ell(\gamma_n) \to d(x, y)$. Parametrize the curves proportional to arclength, and conclude that the lengths of the tangent vectors to γ_n are uniformly bounded. Apply the Arzela-Ascoli theorem to conclude that the γ_n converge to a C^1 curve γ with $\ell(\gamma) = d(x, y)$. Since γ is a geodesic (why?), it must satisfy the Euler-Lagrange equation (2.52). This implies γ is smooth.*

This shortest geodesic is called a *minimal geodesic*. By Exercise 22, the following definition makes sense.

Definition: *Let M be compact. For $x \in M$ the* exponential map $\exp_x : T_x M \to M$ *is defined by*

$$\exp_x(v) = \gamma_v(1).$$

For a noncompact manifold such as $\mathbf{R}^2 - \{(0,0)\}$, the exponential map is defined only on some neighborhood $B_\epsilon(0)$ of $0 \in T_x M$ for each x.

Exercise 24: *(i) Consider S^1 embedded in $\mathbf{R}^2$ in the usual fashion. We have the isomorphism $\mathbf{R} \cong T_{(1,0)} S^1$ given by $\theta \mapsto i\theta$, where we identify $T_{(1,0)} S^1$ with the y axis. Show that $\exp_{(1,0)}(i\theta) = e^{i\theta}$.*

(ii) Show that the exponential map on the standard n-sphere maps radial lines in the tangent space at the north pole to great circles on the sphere.

The key property of the exponential map is that it maps $B_\epsilon(0)$ diffeomorphically onto a neighborhood of x, at least for ϵ small.

Theorem 2.53 *For ϵ small enough, $\exp_x$ restricted to $B_\epsilon(0)$ is a diffeomorphism onto its image, for all $x \in M$. The radius $\epsilon = \epsilon(x)$ can be chosen to be a smooth function of x.*

For example, if $M = S^n$ with the standard metric, then $\exp_x$ maps $B_\pi(0)$ diffeomorphically onto S^n minus x's antipode, but for any $\epsilon > \pi$, the exponential map is no longer a diffeomorphism.

PROOF. The tangent bundle TM is a smooth manifold of dimension twice dim M. In fact, over a chart neighborhood U with coordinates $(x^1, \ldots, x^n)$, TM has coordinates $(x^1, \ldots, x^n, t^1, \ldots, t^n)$, where a vector $v \in T_x M$ is given by $v = t^i \partial_{x^i}$. Thus locally TM looks like the space of initial conditions for the ODE (2.52). The standard existence theorem for ODEs depending on parameters and initial conditions shows that there exists a constant ϵ such that the exponential map is defined on $B_\epsilon(0) \subset T_x M$, for all $x \in U$.

If we set V to be the union of these $B_\epsilon(0)$, the ODE existence theorem guarantees that the map $\exp : V \to M, (x, v) \mapsto \exp_x(v)$, is smooth, and in particular for fixed x the map $\exp_x : B_\epsilon(0) \subset T_x M \to M$ is smooth. We now compute the differential $d(\exp_x)_0 : T_0 T_x M \to T_x M$. Since $T_0 T_x M$ is canonically isomorphic to $T_x M$, we have

$$d(\exp_x)_0(v) = \left.\frac{d}{d\alpha}\right|_{\alpha=0} \exp_x(\alpha v) = \left.\frac{d}{d\alpha}\right|_{\alpha=0} \gamma_{\alpha v}(1).$$

It is easy to check that $\gamma_{\alpha v}(t)$ and $\gamma_v(\alpha t)$ both satisfy (2.52) with the same initial condition. By the uniqueness of solutions of ODEs, $\gamma_{\alpha v}(t) = \gamma_v(\alpha t)$. Thus

$$d(\exp_x)_0(v) = \left.\frac{d}{d\alpha}\right|_{\alpha=0} \gamma_v(\alpha t) = v.$$

Thus the differential of the exponential map at $0 \in T_x M$ is the identity map, so by the inverse function theorem a neighborhood $B_{\epsilon(x)}(0)$ maps diffeomorphically to a neighborhood of x. Moreover, the proof of the inverse function theorem shows that the radius of the domain neighborhood depends smoothly on the differential map, so we may choose $\epsilon(x)$ to be a smooth function of $x \in U$. A partition of unity argument then shows that $\epsilon(x)$ can be chosen to be a smooth function on all of M.

We can now show that for $t < t_0 = t_0(x, v)$, the geodesic $\gamma_v(t)$ is the shortest path between x and $\gamma_v(t)$. The argument we give works for compact manifolds, but can be modified for noncompact manifolds (see [46, §10]). Fix t_1, t with $t_1 < t$ so that $\gamma_v(t)$ lies in the Riemannian normal coordinate chart at x for all $v \in T_x M, |v| = 1$. By Exercise 23, for fixed v there is a shortest path γ from x to $\gamma_v(t_1) = y$. This path is a critical path for length, and so satisfies the geodesic equation. By the uniqueness of solutions to the geodesic equation, we must have $\gamma(s) = \gamma_w(s)$ for some $w \in T_x M, |w| = 1$ (after reparametrizing γ by arclength). If $v \neq w$, then $\gamma_w(s) \neq y$ for all $s \leq t$. Since $\gamma_w(s_0) = y$ for some $s_0 > t$, the length along γ from x to y is at least t. Since the length from x to y along $\gamma_v(s)$ is $t_1 < t$, we get a contradiction. Thus $\gamma_v(t)$ is the minimal geodesic from x to y.

While an arbitrary manifold has no one set of coordinate charts better than any other, a Riemannian manifold has the special set of charts $\{\exp_x : B_\epsilon(0) \to M : x \in M\}$ and corresponding coordinates, called *Riemannian normal coordinates*, given by choosing orthonormal coordinates on $T_x M$. A choice of polar coordinates $(r, \theta) = (r, \theta^1, \ldots, \theta^{n-1})$ on $T_x M$ then gives *Riemannian polar coordinates* on a neighborhood of x via the exponential map. Note that the ∂_r coordinate is the tangent vector to the unit speed geodesics emanating from x.

Exercise 25: *Show that $\Gamma_{ij}^k = 0$ at the center point x when computed in Riemannian normal coordinates. Conclude that $\partial_i g_{jk} = 0$ at x. Hint: In these coordinates, straight lines through the origin map to geodesics, so in these coordinates the Euler-Lagrange equation becomes*

$$\Gamma_{ij}^k \frac{d\gamma^i}{dt} \frac{d\gamma^j}{dt} = 0$$

for all k. At x the $d\gamma^i/dt, d\gamma^j/dt$ are arbitrary, which forces $\Gamma_{ij}^k(x) = 0$. Combining various Γ_{ij}^k gives $\partial_i g_{jk}(x) = 0$.

The following basic result in the study of geodesics will be used in the next chapter.

Theorem 2.54 (Gauss' Lemma) $\langle \partial_r, \partial_{\theta^i} \rangle = 0$ *for* $i = 1, \ldots, n-1$.

PROOF. Let ∂_i denote ∂_{θ_i}, and let ∇_i, ∇_r denote $\nabla_{\partial_i}, \nabla_{\partial_r}$, respectively. We compute how $\langle \partial_r, \partial_i \rangle$ changes as we move out a geodesic radiating from x. We have
$$\partial_r \langle \partial_r, \partial_i \rangle = \langle \nabla_r \partial_r, \partial_i \rangle + \langle \partial_r, \nabla_r \partial_i \rangle.$$
The first term on the right hand side is zero since the radial curve with tangent vector ∂_r is a geodesic. By the standard confusing abuse of notation, we have $\partial_r = d(\exp_x)(\partial_r), \partial_i = d(\exp_x)(\partial_i)$, where ∂_r, ∂_i in the domain of $d\exp_x$ denote the standard polar coordinates on $T_x M$ (with respect to its Riemannian inner product), and ∂_r, ∂_i in the range denote tangent vectors to the Riemannian polar coordinates on a neighborhood of x in M. Thus
$$\nabla_r \partial_i - \nabla_i \partial_r = [\partial_r, \partial_i] = [d\exp_x(\partial_r), d\exp_x(\partial_i)] = d\exp_x([\partial_r, \partial_i]) = 0,$$
since ∂_r, ∂_i are tangent vectors to coordinates in $T_x M$. Thus we obtain
$$\partial_r \langle \partial_r, \partial_i \rangle = \langle \partial_r, \nabla_i \partial_r \rangle = \frac{1}{2} \partial_i \langle \partial_r, \partial_r \rangle = 0,$$
since $\langle \partial_r, \partial_r \rangle = 1$.

In particular, $\langle \partial_r, \partial_i \rangle$ is independent of r. However, in $T_x M$ the length of ∂_i is proportional to r, and hence so is the length of $\partial_i = d(\exp_x)(\partial_i)$ on M. This forces $\langle \partial_r, \partial_i \rangle$ to be proportional to r. The only way this is possible is to have $\langle \partial_r, \partial_i \rangle = 0$.

There are many connections between geodesics and curvature on a Riemannian manifold. In particular, the rate at which geodesics emanating from a point spread apart is controlled by the curvature. Heuristically, this seems reasonable, as geodesics are controlled by an ODE involving one derivative of the metric, and so the variation of geodesics involves two metric derivatives. On the other hand, the curvature, which controls the geometry, involves two metric derivatives, and hence should measure these variations. These remarks are made precise in the study of Jacobi fields, which are behind the proofs of Theorems 2.21, 2.22, and which are discussed in the next section.

For our purposes, we will state without proof another relation between geodesics and curvature, whose demonstration again involves Jacobi fields. Fix $x \in M$ and pick orthonormal coordinates for $T_x M$. For points $y \in T_x M$ with $y \approx 0$, the metric $\exp_x^* g$ is defined at y. The following theorem, due to Cartan, is in [4, Prop. E.III.7].

Theorem 2.55 *The metric $\exp_x^* g$ has the Taylor expansion*

$$g_{ij}(y) = \mathrm{Id} + {}^2 A_{ijkl} y^k y^l + {}^3 A_{ijklm} y^k y^l y^m + \dots$$

with ${}^k A_{ij\dots}$ universal polynomials in the components of the curvature tensor $R_{ijkl} = R_{ijkl}(x)$ and its first $k-2$ covariant derivatives $\nabla R, \nabla^2 R = \nabla \nabla R, \dots$.

Recall that ∇R is the tensor with components given by $R_{ijkl;m}$ as in Exercise 13. The term "universal" means that the coefficients of the polynomials depend only on the dimension of the manifold. In this theorem, we assume that the components of the curvature tensor are computed in Riemannian normal coordinates. This theorem shows that if we know the curvature tensor in a neighborhood of a point, and if we know the collection of radial geodesics at the point (so that we know what the normal coordinate chart is), then the metric at that point is determined, at least for real analytic metrics. This is a technical formulation of Riemann's intuition that the curvature determines the metric; another formulation for smooth metrics is in [64, Vol. II].

2.4 The Laplacian in Exponential Coordinates

This technical section gives a formula for the Laplacian in exponential polar coordinates which will be used in the next chapter.

Fix $x \in M$ and choose a neighborhood U of x which is the diffeomorphic image of a neighborhood of $0 \in T_x M$ under the exponential map. Let $\{X_i = X_i(y)\}$ be an orthonormal frame of $T_y M$ for $y \in U$, and let $\{\theta^i\}$ be the dual frame for $T_y^* M$. Differentiation of functions in the direction X_i will be denoted by ∇_{X_i} or just ∇_i in agreement with Exercise 13(iv). By Lemma 2.39, the Laplacian on functions is given by

$$
\begin{aligned}
\delta d &= -a_i \nabla_i a_j^* \nabla_j = -a_i a_j^* \nabla_i \nabla_j - a_i (\nabla_i \theta^j) \nabla_j \\
&= -\{a_i, a_j^*\} \nabla_i \nabla_j + a_j^* a_i \nabla_i \nabla_j - a_i (\nabla_i \theta^j) \nabla_j \\
&= -\nabla_i \nabla_i - a_i (\nabla_i \theta^j) \nabla_j,
\end{aligned}
$$

where we sum over i, j, and note that $a_i = 0$ on functions. Writing $\nabla_i \theta^j = -\Gamma^j_{ik} \theta^k$ (the minus sign is explained in Exercise 13), we have

$$a_i(\nabla_i \theta^j)\nabla_j = -a_i \Gamma^j_{ik} \theta^k \nabla_j = -\Gamma^j_{ii} \nabla_j = -\nabla_{\nabla_{X_i} X_i},$$

since $a_i \theta^k = \delta^k_i$. Thus we obtain

$$\Delta = -\nabla_i \nabla_i + \nabla_{\nabla_{X_i} X_i}. \tag{2.56}$$

Now fix r small enough and let S be the sphere in M consisting of all points of distance r from p. We want to relate Δ to Δ_S, the Laplacian on S with respect to the induced metric.

Exercise 26: *Let N be a submanifold of a Riemannian manifold M, and give N the induced metric. Show that the Levi-Civita connection $\bar{\nabla}$ on N is given by*

$$\bar{\nabla}_X Y = P_N \nabla_X Y,$$

where X, Y are tangent to N and P_N is the orthogonal projection of $T_x M$ to $T_x N$ (see Exercise 11).

Thus the Levi-Civita connection on S is given by $\bar{\nabla}_X Y = \nabla_X Y - \langle \nabla_X Y, T \rangle T$, where T is the unit tangent vector field to geodesics radiating from p; we may assume that $T = X_n$. Set $h = \langle \nabla_{X_j} X_j, T \rangle$; the sum only goes from 1 to $n-1$, since $\nabla_T T = 0$. Then

$$\begin{aligned}
\Delta_S &= -\nabla_j \nabla_j + \nabla_{\nabla_{X_j} X_j - \langle \nabla_{X_j} X_j, T \rangle} \\
&= -\nabla_j \nabla_j + \nabla_{\nabla_{X_j} X_j} - h \cdot \nabla_T.
\end{aligned}$$

Plugging this last equation into (2.56) gives

$$\Delta = -\nabla_T \nabla_T + \Delta_S + h \cdot \nabla_T. \tag{2.57}$$

In order to simplify the h term, we need to study variations of geodesics. Recall that a geodesic $\gamma(r)$ can be written as $\gamma(r) = \exp(r\gamma'(0))$ for small r. Thus given $\eta \in T_{\gamma(0)}M$, for small r, s the set $\Phi(r, s) = \exp[r(\gamma'(0) + s\eta)]$ describes a two parameter family of geodesics radiating from $\gamma(0)$ with $\Phi(r, 0) = \gamma(0)$. We assume that r, s are small enough so that Φ is injective except when $r = 0$. Let $T = d\Phi(\partial_r), X = d\Phi(\partial_s)$. In particular, $X(r, s)$ measures the spreading of the geodesics $\Phi(r, s)$ as s varies, so X is the variational vector field of the family Φ.

Now

$$\nabla_X T - \nabla_T X = [X, T] = [d\Phi(\partial_r), d\Phi(\partial_s)] = d\Phi([\partial_r, \partial_s]) = 0, \tag{2.58}$$

so by Theorem 2.25

$$\nabla_T \nabla_T X = \nabla_T \nabla_X T = \nabla_X \nabla_T T + R(X, T)T.$$

Since $\nabla_T T = 0$, the variation vector field satisfies the *Jacobi equation*

$$\nabla_T \nabla_T X + R(T, X)T = 0. \tag{2.59}$$

If we consider the Jacobi equation equation restricted to $\gamma(r)$, it becomes a second order ODE in r. Its solutions, which are vector fields defined along γ called *Jacobi fields*, form a vector space of dimension $2n$ ($n = \dim M$), determined by the initial values $X(0), \nabla_T X(0)$.

Note that $X = T$ and $X = rT$ are always Jacobi fields, as both terms in the Jacobi equation are zero. We claim that any Jacobi field can be uniquely written as $X = aT + brT + Y$, where $a, b \in \mathbf{R}$ and Y is a vector field along γ which is pointwise perpendicular to T. For

$$\frac{d^2}{dr^2}\langle X, T\rangle = \langle \nabla_T \nabla_T X, T\rangle = -\langle R(T, X)T, T\rangle = 0,$$

so $\langle X, T\rangle = a + br$. Now set $Y(r) = X(r) - aT - brT$.

We now relate h to $D = D(r) = \det(d\exp_p)$, which measures the distortion of the volume element in $T_p M$ under the exponential map (and thus measures the volume growth of geodesic balls centered at p). To make sense of the determinant, let $q = \exp_p(rT(0))$, and consider the map $d(\exp_p)_{rT(0)} : T_p M \cong T_{rT(0)} T_p M \to T_q M$. Take a parallel orthonormal frame $\{e_1, \ldots, e_n = T = \partial_r\}$ of tangent vectors along $\gamma(r)$; the determinant is computed with respect to the bases $\{e_i(0)\}, \{e_i(r)\}$ of $T_p M, T_q M$, respectively. It is independent of the choice of orthonormal basis, as the reader can check that $D = \det(Pd\exp_p)$, where $P : T_q M \to T_p M$ denotes parallel translation along γ.

Now let Z_i be the Jacobi fields with initial conditions $Z_i(0) = 0, \nabla_T Z_i(0) = e_i$. We claim that

$$Z_i = \frac{d}{ds}\bigg|_{s=0} \exp[r(T(0) + se_i(0))]. \tag{2.60}$$

For $r = 0$ at p, so the right hand side of (2.60) is zero. Also, $\nabla_T = \partial_r$ at p, and so

$$\nabla_T \frac{d}{ds}\bigg|_{s=0} \exp[r(T(0) + se_i(0))](0) = \frac{d}{ds}\bigg|_{s=0} \frac{d}{dr}\bigg|_{r=0} \exp[r(T(0) + se_i(0))]$$

$$= \frac{d}{ds}\bigg|_{s=0} (T(0) + se_i(0)) = e_i(0).$$

Thus the two sides of (2.60) are Jacobi fields with the same initial conditions, and so are equal.

By a similar argument, we have $Z_n = rT$. The other Jacobi fields are of the form $Z_i = a_i T + b_i rT + Y_i$, with $Y_i \perp T$. However, it follows easily from the initial conditions for Z_i that $a_i = b_i = 0$, and so $Z_i \perp T$.

Let $A = A(r)$ be the matrix determined by the equation $Ae_i(r) = Z_i(r)$. Since we have identified $T_{rT(0)} T_p M$ with $T_p M$ by translation by $rT(0)$, D is given by the determinant of the matrix whose columns are the vectors

$$\frac{d}{ds}\bigg|_{s=0} \exp[rT(0) + se_i] = \frac{d}{ds}\bigg|_{s=0} \exp[r(T(0) + se_i/r)].$$

For fixed i, this vector field along γ is just $r^{-1}Z_i$ by (2.60). Thus $D = r^{-n}\det(A)$.

Exercise 27: *Let $A(r)$ be a one parameter family of invertible matrices. Show that*

$$\partial_r \det(A) = \det(A) \cdot \operatorname{tr}(A^{-1}\partial_r A).$$

Hint: We may assume that A has complex coefficients. First show the exercise for A diagonal. Then show the exercise for A diagonalizable.

By the exercise, we have

$$
\begin{aligned}
D^{-1}\nabla_T D &= r^n \det{}^{-1}(A)(-nr^{-n-1}\det(A) + r^{-n}\nabla_T \det(A)) \\
&= -\frac{n}{r} + \operatorname{tr}(A^{-1}\nabla_T A).
\end{aligned}
\tag{2.61}
$$

To calculate the last term, we note that for $i < n$,

$$(A^{-1}\nabla_T A)e_i = A^{-1}\nabla_T(Ae_i),$$

since $\nabla_T e_j = 0$. Thus

$$(A^{-1}\nabla_T A)e_i = A^{-1}\nabla_T Z_i = A^{-1}\nabla_{Z_i}T = A^{-1}A\nabla_{e_i}T = \nabla_{e_i}T$$

by (2.58). Moreover, $AT = Ae_n = Z_n = rT$, so

$$(A^{-1}\nabla_T A)e_n = A^{-1}\nabla_T rT = A^{-1}T = r^{-1}T.$$

Combining this with (2.61) gives

$$
\begin{aligned}
D^{-1}\nabla_T D &= -\frac{n}{r} + \langle r^{-1}T, T\rangle + \sum_{i=1}^{n-1}\langle\nabla_{e_i}T, e_i\rangle \\
&= -\left(\frac{n-1}{r}\right) - \sum_{i=1}^{n-1}\langle\nabla_{e_i}e_i, T\rangle \\
&= -\left(\frac{n-1}{r}\right) - h.
\end{aligned}
\tag{2.62}
$$

Here we have used

$$0 = e_i\langle e_i, T\rangle = \langle\nabla_{e_i}e_i, T\rangle + \langle e_i, \nabla_{e_i}T\rangle.$$

Substituting (2.62) into (2.57) gives

Theorem 2.63 *Let r be the radial coordinate in exponential polar coordinates centered at p, let $T = \partial_r$, let $D = \det(d\exp_p)$ and let Δ_S be the Laplacian on the sphere S of constant distance r_0 from p, for r_0 small. Then for $q \in S$ and $f \in C^\infty(M)$, we have*

$$\Delta f(q) = -\nabla_T\nabla_T f(q) + \Delta_S f(q) - \left(\frac{n-1}{r} + D^{-1}\nabla_T D\right)\nabla_T f(q).$$

In particular, if f is a function of r alone, then

$$\Delta f = -\frac{\partial^2}{\partial r^2}f - \left(\frac{n-1}{r} + \frac{\partial_r \det(d\exp_p)}{\det(d\exp_p)}\right)\partial_r f.$$

Chapter 3

The Construction of the Heat Kernel

In this chapter we will construct the heat kernels for the Laplacians on functions and forms on compact manifolds, and so finally complete the proof of the Hodge theorem given in Chapter 1. At the same time we will study the short time behavior of the heat kernels. This short time behavior contains a surprising amount of geometric information, so much so that M. Kac asked in the 1960s whether the spectrum of the Laplacian, which determines the short time behavior, in fact determines the Riemannian metric itself [37]. We'll return to this question in Chapter 5.

The chapter is organized as follows. In §3.1 we assume the existence of the heat kernel for functions and show that the formal expression given in Chapter 1 for the heat kernel is valid. In §3.2 we construct the heat kernel for functions and indicate the modifications necessary to construct the heat kernel for forms. In §3.3 we study the short time behavior of the heat kernel for functions and forms.

Throughout this chapter M will denote a closed connected oriented manifold of dimension n. The material for this chapter is taken from [4] and [56].

3.1 Preliminary Results for the Heat Kernel

In §1.3.2, we derived the formal expression for the heat kernel,

$$e(t, x, y) = \sum_i e^{-\lambda_i t} \phi_i(x) \phi_i(y),$$

where $\{\phi_i\}$ is an orthonormal basis of $L^2(M)$ satisfying $\Delta \phi_i = \lambda_i \phi_i$. We now show that this expression is valid, provided the heat kernel exists.

Proposition 3.1 *Assume there exists* $e(t, x, y) \in C^\infty(\mathbf{R}^+ \times M \times M)$ *satisfying*

$$(\partial_t + \Delta_y)e(t, x, y) = 0,$$

$$\lim_{t \to 0} \int_M e(t, x, y) f(y) \ \mathrm{dvol}_y \ = \ f(x),$$

for all $f \in L^2(M)$. Then we have the pointwise convergence

$$e(t, x, y) = \sum_i e^{-\lambda_i t} \phi_i(x) \phi_i(y).$$

Moreover, $e(t, x, y)$ is the heat kernel.

PROOF. Let $\{\phi_i\}$ be as above. Fix t and x and write $e(t, x, \cdot) = \sum f_i(t, x) \phi_i(\cdot)$, with equality in $L^2(M)$ in the variable y. Thus $f_i(t, x) = \int_M e(t, x, y) \phi_i(y) \ dy$, where we abbreviate dvol_y by dy. Then

$$
\begin{aligned}
\partial_t f_i(t, x) &= \partial_t \int_M e(t, x, y) \phi_i(y) \ dy \\
&= -\int_M \Delta_y e(t, x, y) \cdot \phi_i(y) \ dy \\
&= -\int_M e(t, x, y) \Delta_y \phi_i(y) \ dy \\
&= -\lambda_i \int_M e(t, x, y) \phi_i(y) \ dy \\
&= -\lambda_i f_i(t, x),
\end{aligned}
$$

and so

$$f_i(t, x) = k_i(x) e^{-\lambda_i t}.$$

Express an arbitrary element $f \in L^2(M)$ as $f = \sum a_i \phi_i$. Then

$$
\begin{aligned}
f(x) &= \lim_{t \to 0} \int_M e(t, x, y) f(y) \ dy \\
&= \lim_{t \to 0} \int_M \sum_i e^{-\lambda_i t} k_i(x) \phi_i(y) \sum_j a_j \phi_j(y) \ dy \\
&= \lim_{t \to 0} \sum_i e^{-\lambda_i t} k_i(x) a_i \\
&= \sum_i k_i(x) a_i,
\end{aligned}
$$

which implies $k_i(x) = \phi_i(x)$. Thus

$$e(t, x, y) = \sum_i e^{-\lambda_i t} \phi_i(x) \phi_i(y)$$

in $L^2(M)$ in the y variable for fixed t, x. As a result, there exists a sequence $i_k \to \infty$ such that

$$\sum_{i=0}^{i_k} e^{-\lambda_i t} \phi_i(x) \phi_i(y) \longrightarrow e(t, x, y)$$

pointwise for any t, x and for almost all y.

By Parseval's equality,

$$
\begin{aligned}
\langle e(t/2, x, \cdot), e(t/2, x', \cdot) \rangle &= \sum_i e^{-\frac{\lambda_i t}{2}} \phi_i(x) e^{-\frac{\lambda_i t}{2}} \phi_i(x') \\
&= \sum_i e^{-\lambda_i t} \phi_i(x) \phi_i(x'),
\end{aligned}
$$

and so $\sum_i e^{-\lambda_i t} \phi_i(x) \phi_i(x')$ converges pointwise with limit continuous in t, x, x'. Therefore $\sum_i e^{-\lambda_i t} \phi_i(x) \phi_i(y) \to e(t, x, y)$ pointwise everywhere.

We leave the last statement to the reader.

Exercise 1: *Show that $\sum_i e^{-\lambda_i t} \phi_i(x) \phi_i(y)$ converges to $e(t, x, y)$ in $H_s(M \times M)$, for each $t > 0$ and for all $s \in \mathbf{R}$. Conclude that we can take integrals and derivatives of the sum term by term. Hint: use Proposition 1.38.*

Corollary 3.2 $\displaystyle \sum_i e^{-\lambda_i t} = \mathrm{Tr}(e^{-t\Delta}) = \int_M e(t, x, x)\, dx.$

PROOF. Since $\sum_i e^{-\lambda_i t} \phi_i(x)^2$ converges to $e(t, x, x)$ for any x and each term is nonnegative, we have

$$
\begin{aligned}
\int_M e(t, x, x)\, dx &= \int_M \sum_i e^{-\lambda_i t} \phi_i(x)^2\, dx = \sum_i e^{-\lambda_i t} \int_M \phi_i(x)^2\, dx \\
&= \sum_i e^{-\lambda_i t} = \mathrm{Tr}(e^{-t\Delta}).
\end{aligned}
$$

3.2 Construction of the Heat Kernel

3.2.1 Construction of the Parametrix

In this subsection we will construct a parametrix for the heat kernel, i.e. an approximate solution of the heat equation defined for x, y close.

Lemma 3.3 $\Delta(fg) = (\Delta f)g - 2\langle df, dg \rangle + f\Delta g.$

PROOF. Since the statement is a pointwise equality, we may check it at a point x using Riemannian normal coordinates centered at x. We may certainly arrange $g_{ij}(x) = \delta_{ij}$ by choosing orthonormal coordinates in $T_x M$, and we know that $\partial_i g_{jk}(x) = 0$ by Exercise 25, Chapter 2. Thus

$$
\Delta = -\frac{1}{\sqrt{\det g}} \partial_{x^i}(\sqrt{\det g}\; g^{ij} \partial_{x^j}) = -\sum_{i=1}^n \frac{\partial^2}{(\partial x^i)^2},
$$

and

$$\langle df, dg \rangle = \left\langle \frac{\partial f}{\partial x^i} dx^i, \frac{\partial g}{\partial x^j} dx^j \right\rangle = \frac{\partial f}{\partial x^i} \frac{\partial g}{\partial x^j} g^{ij}$$

$$= \sum_{i=1}^{n} \frac{\partial f}{\partial x^i} \frac{\partial g}{\partial x^j}.$$

The lemma follows immediately.

By Theorem 2.53, there exists $\epsilon > 0$ such that for all $x \in M$, the exponential map $\exp_x$ takes $B_\epsilon(0) \subset T_x M$ diffeomorphically onto a neighborhood V_x of x. For $y \in V_x$, set $r(x, y)$ to be the length of the radial geodesic joining x to y; note that $r(x, y) < \epsilon$. We define a neighborhood of the diagonal in $M \times M$ by $U_\epsilon = \{(x, y) \subset M \times M : y \in V_x, r(x, y) < \epsilon\}$. Then $G(t, x, y) \equiv (4\pi t)^{-n/2} e^{-\frac{r^2(x,y)}{4t}} \in C^\infty(\mathbf{R}^+ \times U_\epsilon)$. We make an educated guess that the solution of the heat equation on U_ϵ is a modification of the Euclidean heat kernel. So fix $k \in \mathbf{Z}^+$ and set

$$S = S_k = S_k(t, x, y) = (4\pi t)^{-\frac{n}{2}} e^{-\frac{r^2(x,y)}{4t}} (u_0(x, y) + \ldots + u_k(x, y)t^k),$$

for unknown functions $u_i \in C^\infty(U_\epsilon)$. We would like to have $(\partial_t + \Delta_y)S = 0$. Now

$$\frac{\partial S}{\partial t} = G \cdot \left(\left(-\frac{n}{2t} + \frac{r^2}{4t^2} \right)(u_0 + \ldots + t^k u_k) \right.$$
$$\left. + (u_1 + 2u_2 t + \ldots + k u_k t^{k-1}) \right), \tag{3.4}$$

and

$$\Delta_y S = (\Delta G)(u_0 + \ldots + u_k t^k) - 2\langle dG, d(u_0 + \ldots + u_k t^k) \rangle$$
$$+ G\Delta(u_0 + \ldots + u_k t^k), \tag{3.5}$$

by Lemma 3.3. Also,

$$\Delta G = -\frac{\partial^2 G}{\partial r^2} - \frac{\partial G}{\partial r} \left(\frac{D'}{D} + \frac{n-1}{r} \right)$$
$$= \left(\frac{n}{2t} - \frac{r^2}{4t^2} \right) G + \frac{r}{2t} \frac{D'}{D} G, \tag{3.6}$$

by Theorem 2.63 of Chapter 2, where D' denotes $\partial_r D$, and

$$\langle dG, d(u_0 + \ldots + u_k t^k) \rangle$$
$$= \left\langle \frac{\partial G}{\partial r} dr + \frac{\partial G}{\partial \theta} d\theta, d(u_0 + \ldots + u_k t^k) \right\rangle$$
$$= \left\langle \frac{\partial G}{\partial r} dr, \frac{\partial u_0}{\partial r} dr + \frac{\partial u_0}{\partial \theta} d\theta + \ldots + t^k \frac{\partial u_k}{\partial r} dr + t^k \frac{\partial u_k}{\partial \theta} d\theta \right\rangle$$

$$
\begin{aligned}
&= \; \left\langle \frac{\partial G}{\partial r} dr, \frac{\partial u_0}{\partial r} dr + \ldots + t^k \frac{\partial u_k}{\partial r} dr \right\rangle \\[2mm]
&= \; \frac{\partial G}{\partial r} \left(\frac{\partial u_0}{\partial r} + \ldots + t^k \frac{\partial u_k}{\partial r} \right) \\[2mm]
&= \; -\frac{r}{2t} \left(\frac{\partial u_0}{\partial r} + \ldots + t^k \frac{\partial u_k}{\partial r} \right) G,
\end{aligned}
\tag{3.7}
$$

where $(\partial G/\partial \theta)d\theta$ has the obvious meaning and we have used Gauss' lemma. Combining (3.4)–(3.7), we get

$$
\begin{aligned}
(\partial_t + \Delta_y)S \;=\; & G \cdot \left(u_1 + \ldots + k t^{k-1} u_k + \frac{r}{2t}\frac{D'}{D}(u_0 + \ldots + t^k u_k) \right. \\[2mm]
& \left. + \frac{r}{t}\left(\frac{\partial u_0}{\partial r} + \ldots + t^k \frac{\partial u_k}{\partial r} \right) + \Delta_y u_0 + \ldots + t^k \Delta_y u_k \right).
\end{aligned}
\tag{3.8}
$$

While it is not possible in general to pick the u_i so that the right hand side of (3.8) is zero, we can make this expression vanish up to the highest power of t, i.e. we can solve $(\partial_t + \Delta)S = (4\pi t)^{-\frac{n}{2}} e^{-r^2(x,y)/4t} t^k \Delta_y u_k(x, y)$ by making all the terms containing $t^{i-(n/2)-1}$ vanish, for $i = 0, 1, \ldots, k$. This leads to the following series of so-called transport equations; the first equation is obtained by setting the coefficient of $t^{-(n/2)-1}$ equal to zero, and the second by doing the same for $t^{i-(n/2)-1}$.

$$
r\frac{\partial u_0}{\partial r} + \frac{r}{2}\frac{D'}{D}u_0 \;=\; 0,
\tag{3.9}
$$

$$
r\frac{\partial u_i}{\partial r} + \left(\frac{r}{2}\frac{D'}{D} + i \right)u_i + \Delta_y u_{i-1} \;=\; 0, \quad i = 1, \ldots, k.
\tag{3.10}
$$

(3.9) reduces to

$$
\frac{\partial \ln u_0}{\partial r} = -\frac{1}{2}\frac{\partial}{\partial r} \ln D,
$$

so $u_0 = kD^{-\frac{1}{2}}$, where $k = k(\theta)$. Since we want u_0 defined at $r = 0$, we must set k equal to a constant. Setting $k = 1$, we get

$$
u_0(x, y) = \frac{1}{\sqrt{D(\exp_x^{-1}(y))}},
\tag{3.11}
$$

and in particular

$$
u_0(x, x) = 1.
$$

Exercise 2: *What goes wrong with the proof below if we pick k to be a constant other than one?*

To find the other u_i, we first solve a simpler version of (3.10),

$$
r\frac{\partial u_i}{\partial r} + \left(\frac{r}{2}\frac{D'}{D} + i \right)u_i = 0.
$$

This has the solution $kr^{-i}D^{-1/2}$, with $k = k(\theta)$ arbitrary. Operating with hindsight, we assume that u_i has the form $u_i = kr^{-i}D^{-1/2}$, but now with $k = k(r)$. Plugging this expression into (3.10) gives

$$\frac{\partial k}{\partial r} = -D^{1/2}(\Delta u_{i-1})r^{i-1}.$$

Let $x(s)$ be the unit speed geodesic from x to y, $s \in [0, r]$. Δu_{i-1} is a function of r along this geodesic, and so the last equation can be solved by an r integration. Substituting this solution into the assumed form for u_i, we obtain

$$u_i(x,y) = -r^{-i}(x,y)D^{-\frac{1}{2}}(y)\int_0^r D^{\frac{1}{2}}(x(s))\Delta_y u_{i-1}(x(s),y)s^{i-1}ds. \qquad (3.12)$$

In summary, for u_i defined inductively by (3.11), (3.12), we have

$$(\partial_t + \Delta_y)S_k = (4\pi t)^{-\frac{n}{2}}e^{-\frac{r^2}{4t}}t^k\Delta u_k. \qquad (3.13)$$

An induction argument shows that $u_i \in C^\infty(U_\epsilon)$. We now extend S to a function on all of $M \times M$. Pick a bump function $\eta \in C^\infty(M \times M)$ with $\eta(x,y) \in [0,1]$, $\eta \equiv 0$ on $M \times M - U_\epsilon$, and $\eta \equiv 1$ on $U_{\epsilon/2}$. Set $H_k \equiv \eta S_k \in C^\infty(\mathbf{R}^+ \times M \times M)$.

In the next definition, we use the notation $\mathbf{R}^0 \equiv \mathbf{R}^+ \cup \{0\}$.

Definition: *A parametrix for the heat operator $\partial_t + \Delta_y$ is a function $H(t,x,y) \in C^\infty(\mathbf{R}^+ \times M \times M)$ such that (a) $(\partial_t + \Delta_y)H \in C^0(\mathbf{R}^0 \times M \times M)$ and (b) $\lim_{t\to 0}\int_M H(t,x,y)f(y)dy = f(x)$.*

Lemma 3.14 H_k *is a parametrix if $k > n/2$. In fact $(\partial_t + \Delta_y)H_k \in C^l(\mathbf{R}^0 \times M \times M)$ if $k > l + \frac{n}{2}$.*

PROOF. We leave the second statement to the reader. To see that $(\partial_t + \Delta_y)H$ extends to $t = 0$, first note that $H \equiv 0$ on $\mathbf{R}^+ \times (M \times M - U_\epsilon)$, and so $(\partial_t + \Delta_y)H$ trivially extends. On $\mathbf{R}^+ \times U_{\epsilon/2}$,

$$(\partial_t + \Delta_y)H_k = (\partial + \Delta_y)S_k = (4\pi t)^{-\frac{n}{2}}t^k e^{-\frac{r^2}{4t}}\Delta u_k \to 0,$$

as $t \to 0$. Thus $(\partial_t + \Delta_y)H$ again extends by zero on this set. Finally, on $U_\epsilon - U_{\epsilon/2}$,

$$\begin{aligned}(\partial_t + \Delta_y)H_k &= \eta(\partial_t + \Delta_y)S_k - 2\langle d\eta, dS_k\rangle + (\Delta_y\eta)S_k \\ &= (4\pi t)^{-\frac{n}{2}}e^{-\frac{r^2}{4t}}\phi(t,x,y),\end{aligned}$$

for some function $\phi(t,x,y) \in C^\infty(\mathbf{R}^+ \times M \times M)$ with at most a pole of order t^{-1} at $t = 0$. Since $r > \frac{\epsilon}{2}$, we may extend $(\partial_t + \Delta_y)H$ by zero.

To finish the proof, we must show that

$$\lim_{t\to 0}\int_M (4\pi t)^{-\frac{n}{2}}e^{-\frac{r^2}{4t}}\eta(x,y)(u_0(x,y) + \ldots + t^k u_k(x,y))f(y)dy = f(x).$$

Now

$$\lim_{t\to 0}\int_M (4\pi t)^{-\frac{n}{2}}\eta(x,y)e^{-\frac{r^2}{4t}}u_i(x,y)f(y)dy$$

$$= \lim_{t\to 0}\int_{B_{\epsilon/2}(x)} (4\pi t)^{-\frac{n}{2}}\eta(x,y)e^{-\frac{r^2}{4t}}u_i(x,y)f(y)dy$$

$$+\lim_{t\to 0}\int_{M-B_{\epsilon/2}(x)} (4\pi t)^{-\frac{n}{2}}\eta(x,y)e^{-\frac{r^2}{4t}}u_i(x,y)f(y)dy.$$

The second integral on the right hand side vanishes as $t \to 0$, since $r > \epsilon/2$. Using the exponential map as a coordinate chart, the first integral on the right hand side becomes an ordinary integral over $\mathbf{R}^n \cong T_x M$:

$$\int_{B_{\epsilon/2}(x)} (4\pi t)^{-\frac{n}{2}}\eta(x,y)e^{-\frac{r^2}{4t}}u_i(x,y)f(y)dy$$

$$= \int_{B_{\epsilon/2}(0)\subset T_x M} (4\pi t)^{-\frac{n}{2}}e^{-\frac{r^2(0,v)}{4t}}u_i(x,\exp_x v)f(\exp_x v)D(v)dv^1\ldots dv^n$$

$$= \int_{T_x M} (4\pi t)^{-\frac{n}{2}}e^{-\frac{r^2(0,v)}{4t}}u_i(x,\exp_x v)f(\exp_x v)D(v)dv^1\ldots dv^n,$$

with u_i extended to be zero off $B_{\epsilon/2}(x)$. Since $(4\pi t)^{-\frac{n}{2}}e^{-\frac{r^2}{4t}}$ is the ordinary heat kernel of $\mathbf{R}^n$, as $t \to 0$ the last integral converges to

$$u_i(x,\exp_x 0)f(\exp_x 0)D(0) = u_i(x,x)f(x).$$

Since $u_0(x,x) = 1$, we see that

$$\lim_{t\to 0}\int_M (4\pi t)^{-\frac{n}{2}}e^{-\frac{r^2}{4t}}\eta(x,y)u_0(x,y)f(y)dy = f(x),$$

and

$$\lim_{t\to 0}\int_M (4\pi t)^{-\frac{n}{2}}e^{-\frac{r^2}{4t}}\eta(x,y)t^i u_i(x,y)f(y)dy = 0,$$

for $i > 0$.

3.2.2   The Heat Kernel for Functions

In this subsection we will finish the construction of the heat kernel for functions. Recall that we have a parametrix $H_k(t,x,y)$ which is a good approximation to the heat kernel for x, y close and t small. To promote this approximation to a full heat kernel, we use the technique of iterating Duhamel's formula. This construction is technical, so we will begin with a general discussion.

Let X, Y be operators on a Hilbert space of functions. We will assume that X, Y have well defined heat operators e^{-tX}, e^{-tY}, i.e. a semigroup of bounded self-adjoint operators satisfying

$$(\partial_t + X)e^{-tX}f = 0, \quad \lim_{t\to 0}e^{-tX}f = f,$$

and similarly for Y. In the following, we will denote expressions such as $e^{-tX}f$ just by e^{-tX}.

Proposition 3.15 (Duhamel's formula) *Provided $e^{-t(X+Y)}$ exists, we have*

$$e^{-t(X+Y)} = e^{-tX} - \int_0^t e^{-(t-s)(X+Y)} Y e^{-sX} \, ds.$$

PROOF. (see [18]) As in (1.26), e^{-tX} is injective and we denote its (possibly unbounded) inverse by e^{tX}. (Since e^{-tX} is injective and self-adjoint, it is surjective.) Set $B(t) = e^{-t(X+Y)}e^{tX}$. Then

$$\frac{dB}{dt} = e^{-t(X+Y)}(-(X+Y))e^{tX} + e^{-t(X+Y)}e^{tX}X = -e^{-t(X+Y)}Ye^{tX},$$

since formally (and rigorously by the spectral theory of unbounded operators) $e^{tX}X = Xe^{tX}$. Thus

$$e^{-t(X+Y)}e^{tX} - \mathrm{Id} = -\int_0^t e^{-s(X+Y)}Ye^{sX} \, ds,$$

and so

$$\begin{aligned} e^{-t(X+Y)} - e^{-tX} &= -\int_0^t e^{-s(X+Y)}Ye^{(s-t)X} \, ds \\ &= -\int_0^t e^{-(t-s)(X+Y)}Ye^{-sX} \, ds. \end{aligned}$$

Given operators $A(t), B(t)$ on our Hilbert space, set

$$A * B = \int_0^t A(t-s)B(s) \, ds.$$

For example, we may rewrite Duhamel's formula as

$$e^{-t(X+Y)} = e^{-tX} - e^{-t(X+Y)} * (Ye^{-tX}).$$

We denote the λ-fold product $A * \ldots * A$ by $A^{*\lambda}$ and set $A^{*1} = A$. It is easy to check that $*$ is associative, so the notation is unambiguous.

Corollary 3.16 *We have*

$$e^{-t(X+Y)} = e^{-tX} + \sum_{j=1}^n (-1)^j b_j + (-1)^{n+1} r_{n+1},$$

where

$$b_n = e^{-tX} * (Ye^{-tX})^{*n}$$

and

$$r_n = e^{-t(X+Y)} * (Ye^{-tX})^{*n}.$$

PROOF. The proof is by induction on n. For $n = 0$, the corollary reduces to Duhamel's formula (with $b_0 = 0$). For the induction step, we apply Duhamel's formula to the $e^{-t(X+Y)}$ term in r_n to get

$$r_n = (e^{-tX} - e^{-t(X+Y)} * (Ye^{-tX})) * (Ye^{-tX})^{*n}$$
$$= b_n - r_{n+1}.$$

The corollary follows.

Given X, Y, we can construct the heat operator for $X + Y$ from the heat operator of X from the corollary:

$$e^{-t(X+Y)} = e^{-tX} + e^{-tX} * \sum_{\lambda=1}^{\infty}(-1)^{\lambda}(Ye^{-tX})^{*\lambda}, \qquad (3.17)$$

provided $r_{n+1} \to 0$ in the Hilbert space. We now adapt this expression formally to our setup. Let $A(t)$ be the operator on $L^2(M, g)$ with kernel $H(t, x, y) = H_k(t, x, y)$. We very dangerously assume that $A(t)$ is like a heat operator, so that there exists an operator X on L^2 such that $(\partial_t + X)A(t) = 0$. Thus $e^{-tX} = A(t)$ and $(\partial_t + X)H = 0$. Set $Y = \Delta - X$ and $K = (\partial_t + \Delta)H$. Note that the kernel of $YA(t)$ is just $Y_x H(t, x, y)$. Extend the Hilbert space $\mathcal{H}$ of functions to the space of time dependent functions $\mathcal{H} \times \mathbf{R}$. Then e^{-tX} acts on this space by

$$e^{-tX}\psi(t, x) = \int_0^t d\theta \int_M H(t - \theta, x, z)\psi(\theta, z)\text{dvol}(z)$$

(see Exercise 5, Chapter 4). Thus if $B(t)$ is any operator with kernel $B(t, x, y)$, $A * B$ has kernel

$$\int_0^t d\theta \int_M H(\theta, x, q)B(t - \theta, q, y) \, dq.$$

Interpreting (3.17) at the level of kernels gives

$$e(t, x, y) = H(t, x, y) + [H * \sum_{\lambda=1}^{\infty}(-1)^{\lambda}((\Delta - X)H)^{*\lambda}](t, x, y)$$

$$= H(t, x, y) + [H * \sum_{\lambda=1}^{\infty}(-1)^{\lambda}((\partial_t + \Delta)H)^{*\lambda}](t, x, y)$$

$$= H(t, x, y) + [H * \sum_{\lambda=1}^{\infty}(-1)^{\lambda}K^{*\lambda}](t, x, y).$$

This is our formal expression for the heat kernel.

It is remarkable that this formal procedure actually works, as we will now show. Of course, the hard work involves controlling the error term r_n. To begin, for $A, B \in C^0(\mathbf{R}^0 \times M \times M)$, note that

$$(A * B)(t, x, y) = \int_0^t d\theta \int_M A(\theta, x, q)B(t - \theta, q, y)dq \in C^0(\mathbf{R}^0 \times M \times M).$$

Lemma 3.18 *Set $K_k = (\partial_t + \Delta_y)H_k$. Then $Q_k = \sum_{\lambda=1}^{\infty}(-1)^{\lambda+1}K_k^{*\lambda}$ exists and is in $C^l(\mathbf{R}^0 \times M \times M)$ if $k > l + (n/2)$. Moreover, given $T > 0$, there exists $C = C(T)$ so that $|Q_k(t,x,y)| \leq C \cdot t^{k-\frac{n}{2}}$ for all $t \in [0,T]$.*

PROOF. Writing $K_k = (\partial_t + \Delta_y)(\eta S_k)$ and performing the differentiation, we easily get

$$|K_k| \leq A(T)t^{k-(n/2)} \leq A(T)T^{k-(n/2)} \equiv B,$$

for some constants $A = A(T), B$. (The reader should check that, as in the proof of Lemma 3.14, $|K_k|$ does not blow up as $t \to 0$, as $d\eta$ and $\Delta\eta$ have support bounded away from the diagonal in $M \times M$.)

We claim that

$$|K_k^{*\lambda}(t,x,y)| \leq \frac{AB^{\lambda-1}V^{\lambda-1}t^{k-(n/2)+\lambda-1}}{(k-\frac{n}{2}+1)(k-\frac{n}{2}+2)\ldots(k-\frac{n}{2}+\lambda-1)}, \tag{3.19}$$

where $V = \mathrm{vol}(M)$. The case $\lambda = 1$ has just been done, provided we readjust the definition of A. Assuming the claim for $\lambda - 1$, we have

$$\begin{aligned}
|K_k^{*\lambda}| &\leq \int_0^t d\theta \int_M |K_k^{*(\lambda-1)}(\theta,x,q)| \, |K_k(t-\theta,q,y)|dq \\
&\leq \int_0^t d\theta \int_M \frac{AB^{\lambda-2}V^{\lambda-2}\theta^{k-(n/2)+\lambda-2}}{(k-\frac{n}{2}+1)\ldots(k-\frac{n}{2}+\lambda-2)} \cdot B \\
&= \frac{AB^{\lambda-1}V^{\lambda-2}V}{(k-\frac{n}{2}+1)\ldots(k-\frac{n}{2}+\lambda-2)}\int_0^t \theta^{k-(n/2)+\lambda-2}d\theta.
\end{aligned}$$

Evaluating the last integral finishes the claim.

The right hand side of (3.19) is bounded by a constant times

$$\frac{(BVt)^{\lambda-1}t^{k-(n/2)}}{\Gamma(k-\frac{n}{2}+\lambda-2)}, \tag{3.20}$$

so the ratio test shows that $\sum_{\lambda=1}^{\infty}|K_k^{*\lambda}|$ converges. This easily implies that $\sum_{\lambda=1}^{\infty}(-1)^{\lambda+1}K_k^{*\lambda}$ converges to a continuous function in t,x,y if $k > \frac{n}{2}$. The estimate (3.20) also implies $|Q_k| \leq C \cdot t^{k-(n/2)}$ for some constant C. We leave to the reader similar estimates on the derivatives of $K_k^{*\lambda}$ which show that $Q_k \in C^l$ if $k > l + \frac{n}{2}$.

Lemma 3.21 *(i) If $P \in C^0(\mathbf{R}^0 \times M \times M)$, then $P * H_k \in C^l(\mathbf{R}^+ \times M \times M)$, if $k > l + \frac{n}{2}$.*
  (ii) $(\partial_t + \Delta)(P * H_k) = P + P * K_k$ if $k > 2 + \frac{n}{2}$.

PROOF. (i) is left for the reader; the problem is that H_k blows up at zero. For (ii), by the Leibniz rule,

$$(\partial_t + \Delta_y)(P * H_k) = (\partial_t + \Delta_y)\int_0^t d\theta \int_M P(\theta,x,q)H_k(t-\theta,q,y)\,dq$$

$$= \lim_{s \to t} \int_M P(s, x, q) H_k(t - s, q, y) \, dq$$

$$+ \int_0^t d\theta \int_M P(\theta, x, q) \cdot \partial_t H_k(t - \theta, q, y) \, dq$$

$$+ \int_0^t d\theta \int_M P(\theta, x, q) \cdot \Delta_y H_k(t - \theta, q, y) \, dq$$

$$= P(t, x, y)$$

$$+ \int_0^t d\theta \int_M P(\theta, x, q) \cdot (\partial_t + \Delta_y) H_k(t - \theta, q, y) \, dq$$

$$= P + P * K_k.$$

We now finally produce the heat kernel for functions.

Theorem 3.22 *Set $e(t, x, y) = H_k(t, x, y) - Q_k * H_k(t, x, y)$. Then $e(t, x, y) \in C^\infty(\mathbf{R}^+ \times M \times M)$ is independent of k if $k > 2 + \frac{n}{2}$ and is the heat kernel.*

PROOF. By our hypothesis on k, $e(t, x, y)$ is C^2. The proof of Proposition 3.1 in fact only uses this amount of differentiability, so to prove the theorem, it suffices to show that $e(t, x, y)$ satisfies the hypotheses of the proposition. By Lemma 3.21,

$$
\begin{aligned}
(\partial_t + \Delta_y) e(t, x, y) &= (\partial_t + \Delta_y)(H_k - Q_k * H_k) \\
&= K_k - Q_k - Q_k * K_k \\
&= K_k - \sum_{\lambda=1}^{\infty} (-1)^{\lambda+1} K_k^{*\lambda} - \sum_{\lambda=1}^{\infty} (-1)^{\lambda+1} K_k^{*\lambda} * K_k \\
&= 0.
\end{aligned}
$$

We also have

$$
\begin{aligned}
\lim_{t \to 0} \int_M e(t, x, y) f(y) dy &= \lim_{t \to 0} \left(\int_M H_k(t, x, y) f(y) dy \right. \\
&\qquad \left. - \int_M (Q_k * H_k)(t, x, y) f(y) dy \right) \\
&= f(x) - \lim_{t \to 0} \int_M (Q_k * H_k)(t, x, y) f(y) dy.
\end{aligned}
$$

Now $R_k \equiv Q_k / (t^{k - (n/2)})$ is bounded for all t in some finite interval, so by Lemma 3.14, for $k > \frac{n}{2}$,

$$\lim_{t \to 0} \int_M (Q_k * H_k)(t, x, y) f(y) dy = \lim_{t \to 0} t^{k - (n/2)} \int_M (R_k * H_k)(t, x, y) f(y) dy = 0.$$

Thus $e(t, x, y)$ is the heat kernel, and by the uniqueness of the heat kernel must be independent of k. Finally, $H_k - Q_k * H_k$ is in $C^{k - (n/2)}(\mathbf{R}^+ \times M \times M)$ for any k, i.e. $e(t, x, y) \in C^\infty(\mathbf{R}^+ \times M \times M)$.

Exercise 3: *Given $\epsilon > 0, T > 0$ and $x \in M$, show that $e(t,x,y) = O(t)$ for $t \in (0,T]$ and $y \in M \setminus B_\epsilon(x)$. Hint: Show that this estimate is true for K_k and hence for K_k^λ and Q_k. The integrals for K_k^λ can be handled as in Lemma 4.23.*

The construction of the heat kernel for forms follows along the same lines. First, we prove a product rule for Δ^k on forms as in Lemma 3.3.

Exercise 4: *On a Riemannian manifold, define an operation $\langle \ , \ \rangle_x : T_x^* M \otimes (\Lambda^k T_x^* M \otimes T_x^* M) \to \Lambda^k T_x^* M$ as the linear extension of $\alpha \otimes (\beta_1 \otimes \beta_2) \mapsto \langle \alpha, \beta_2 \rangle \beta_1$. This is a type of contraction. Show that for a function f and a k-form ω, we have*

$$\Delta^k(f\omega) = (\Delta^0 f)\omega - 2\langle df, \nabla \omega \rangle + f\Delta^k \omega.$$

Hint: By Taylor's theorem and Exercise 25, Chapter 2, we know that in Riemannian normal coordinates centered at x, $g_{ij}(y) = \delta_{ij} + O(r^2(x,y))$ for y close to x. This implies that to prove the exercise at x, we may assume that the metric is flat near x – i.e. $g_{ij}(y) = \delta_{ij}$ (why?).

Continuing with the construction, we must alter (3.12) by replacing the term $\Delta_y u_{i-1}(x(s), y)$ with $||_s \Delta_y u_{i-1}(x(s), y)$, where $||_s$ denotes parallel translation along the geodesic $x(\cdot)$ from $x(s)$ to x. We then define S_k, H_k, K_k as for functions. To compute the growth of K_k^λ, we use the pointwise Hodge norms $|\omega|_x^2 = \langle \omega, \omega \rangle_x$ on $\omega \in \Lambda^k T^* M$. The analogue of Lemma 3.18 then holds. The rest of the construction is straightforward. Complete details are in [56].

3.3 The Asymptotics of the Heat Kernel

In accordance with §1.1, we write $A(t) \sim \sum_{k=k_0}^{\infty} b_k t^k$ if for all $N \geq k_0$,

$$\lim_{t \to 0} \frac{A(t) - \sum_{k=k_0}^{N} b_k t^k}{t^N} = 0.$$

Note that $\sum b_k t^k$, which is called the asymptotic expansion of $A(t)$, need not converge for any value of t.

Proposition 3.23 *$e(t,x,x)$ has the asymptotic expansion*

$$e(t,x,x) \sim (4\pi t)^{-n/2} \sum_{k=0}^{\infty} u_k(x,x) t^k.$$

PROOF. We know

$$e(t,x,x) = (H_k - Q_k * H_k)(t,x,x),$$

for any $k \gg 0$. Since $|Q_k| \leq C \cdot t^{k-(n/2)}$, we have $(4\pi t)^{n/2}|Q_k * H_k| \leq C \cdot t^{k+1}$. As $(4\pi t)^{n/2} H_k(t,x,x) = u_0(x,x) + u_1(x,x)t + \ldots + u_k(x,x)t^k$, we see that

$$(4\pi t)^{n/2} e(t,x,x) = u_0(x,x) + \ldots + u_k(x,x)t^k + R(t,x),$$

where $\lim_{t\to 0}(R(t,x)/t^k) = 0$.

Exercise 5: *Show that for x, y close enough, we have*

$$\lim_{t\to 0} \frac{e(t,x,y) - \frac{e^{-r^2(x,y)/4t}}{(4\pi)^{n/2}} \sum_{k=0}^{N} u_k(x,y)t^{k-(n/2)}}{t^N} = 0.$$

Conclude as in Exercise 3 that for $x \neq y$, $e(t,x,y) \sim 0$.
     (ii) It is unclear from the construction of $u_i(x,y)$ whether $u_i(x,y) = u_i(y,x)$. Does this symmetry follow from the symmetry of $e(t,x,y)$?

This gives the important asymptotic expansion for the trace of the heat kernel:

Theorem 3.24 *Let $\{\lambda_i\}$ be the spectrum of the Laplacian on functions on (M, g). Then*

$$\sum_i e^{-\lambda_i t} \sim (4\pi t)^{-n/2} \sum_{k=0}^{\infty} a_k t^k,$$

with

$$a_k = \int_M u_k(x,x)\, \mathrm{dvol}(x).$$

PROOF.

$$\sum_i e^{-\lambda_i t} = \int_M e(t,x,x) \quad \sim \quad \int_M (4\pi t)^{-n/2} \sum_k u_k(x,x)t^k\, \mathrm{dvol}$$

$$\sim \quad (4\pi t)^{-n/2} \sum_k \left(\int_M u_k(x,x)\, \mathrm{dvol} \right) t^k.$$

We can now give the first example of how the spectrum of the Laplacian on functions determines interesting geometric quantites. We say that two Riemannian manifolds are *isospectral* if the eigenvalues of their Laplacians on functions, counted with multiplicities, coincide.

Corollary 3.25 *Let M and N be compact isospectral Riemannian manifolds. Then M and N have the same dimension and the same volume.*

PROOF. If $\{\lambda_i\}$ denotes the spectrum for M and N, and if $\dim M = m, \dim N = n$, we have

$$(4\pi t)^{-m/2} \sum_{k=0}^{\infty} \left(\int_M u_k^M(x,x) \right) t^k \quad \sim \quad \sum_i e^{-\lambda_i t}$$

$$\sim \quad (4\pi t)^{-n/2} \sum_{k=0}^{\infty} \left(\int_N u_k^N(x,x) \right) t^k.$$

This immediately implies $m = n$. Thus

$$(4\pi t)^{-m/2} [\int_M u_0^M(x,x) - \int_N u_0^N(x,x)]$$

$$\sim \quad (4\pi t)^{-m/2} \sum_{k=1}^{\infty} \left(\int_M u_k^M(x,x) - \int_N u_k^N(x,x) \right) t^k.$$

Again, this implies

$$\int_M u_0^M(x,x) = \int_N u_0^N(x,x),$$

and by iterating this argument,

$$\int_M u_k^M(x,x) = \int_N u_k^N(x,x),$$

for all k. In particular, since $u_0(x,x) = 1$, we obtain $\mathrm{vol}(M) = \mathrm{vol}(N)$.

Exercise 6: *Show that for the circle of circumference 2π, we have $u_i(x,x) \equiv 0$ for $i > 0$. This recovers Jacobi's formula (Theorem 1.12). Conclude that any manifold isospectral to this circle must be isometric to this circle. Hint: From §1.1, the heat kernel on S^1 is asymptotic to the heat kernel on $\mathbf{R}$ restricted to an interval of length 2π.*

The proof shows that in fact there are an infinite sequence of obstructions to two manifolds being isospectral, namely the $\int_M u_k$. Since the first integral contains basic geometric information, it is natural to investigate the other integrals. We will now begin to compute $u_1(x,x)$. Recall that $R_x, \nabla R_x, \nabla^2 R_x, \ldots$ denote the covariant derivatives of the curvature tensor at x, and that a polynomial P in the curvature and its covariant derivatives is called universal if its coefficients depend only on the dimension of M. We use the notation $P(R_x, \nabla R_x, \ldots, \nabla^k R_x)$ to denote a polynomial in the components of the curvature tensor and its covariant derivatives, computed in a Riemannian normal coordinate chart at x.

Lemma 3.26 *On an n-manifold,*

$$u_1(x,x) = P_1^n(R_x),$$
$$u_i(x,x) = P_i^n(R_x, \nabla R_x, \nabla^2 R_x, \ldots, \nabla^{2i-4} R_x), \quad i \geq 2,$$

for some universal polynomials P_i^n.

PROOF. Of course, $u_0(x,x) = 1$, but note further that for y close to x, $u_0(x,y) = \det^{-1/2} g(y)$ has a Taylor expansion in the components of y, with coefficients universal polynomials in $R_x, \nabla R_x, \ldots$, because (i) the g_{ij} have such a Taylor expansion by Theorem 2.55, and (ii) $x^{-1/2}$ has its ordinary Taylor expansion. Assume by induction that $u_{i-1}(x,y)$ has a Taylor expansion in y, with coefficients given by universal polynomials in the curvature and its covariant derivatives, for all $x, y \in M$ close enough. We have the recursion formula

$$u_i(x, y) = -r^{-i}(x, y) \det(y)^{-\frac{1}{2}} \int_0^r \det(x(s))^{\frac{1}{2}} \Delta_y u_{i-1}(x(s), y) s^{i-1} ds,$$

where $r = d(x, y)$ and $x(s)$ is the geodesic from x to $y = x(r)$. Note that if $y = (y^1, \ldots, y^n)$ in Riemannian normal coordinates at x, then $r^2(y) = g_{ij} y^i y^j$ (check this), and so $r^{-i}(y)$ has a Taylor expansion in y, with coefficients given by universal polynomials as above.

Exercise 7: *Show that if $u_{i-1}(x, y)$ satisfies the induction hypothesis for all x, y close, then so does $\Delta_y u_{i-1}(x, y)$. Hint: Use the local expression (1.14) for Δ, and show that g^{ij} has such a Taylor expansion.*

It follows from the exercise and the last equation that $u_i(x, y)$ also satisfies the induction hypothesis. Setting $r = 0$ computes $u_i(x, x)$, which is just the constant term in the Taylor expansion of $u_i(x, y)$, and is thus a universal polynomial in the curvature and its covariant derivatives.

We now determine which curvature terms may occur in the $u_i(x, x)$ by a scaling argument.

Exercise 8: *(i) Pick $\lambda \in \mathbf{R}$ and consider the scaled metric $\lambda^2 g$. Show that $\Delta^{\lambda^2 g} = \lambda^{-2} \Delta^g$, where Δ^g is the Laplacian for the metric g. Conclude that the spectrum of $\Delta^{\lambda^2 g}$ is $\{\lambda^{-2} \lambda_i\}$, where $\{\lambda_i\}$ is the spectrum of Δ^g.*

(ii) Show that if $\{\phi_i\}$ is an orthonormal basis of $L^2(M, g)$ consisting of eigenfunctions of Δ^g, then $\{\lambda^{-n/2} \phi_i\}$ is an orthonormal basis of $L^2(M, \lambda^2 g)$ consisting of eigenfunctions of $\Delta^{\lambda^2 g}$.

(iii) Let R_g, ∇_g^k denote the curvature tensor and the iterated covariant derivative for the metric g, respectively. Show that $\nabla_{\lambda^2 g}^k R_{\lambda^2 g} = \lambda^{-2-k} \nabla_g^k R_g$.

Let $u_i^g(x, x)$ denote the asymptotic coefficients for the heat kernel for the metric g.

Lemma 3.27 *For $\lambda \in \mathbf{R}$, $u_k^{\lambda^2 g}(x, x) = \lambda^{-2k} u_k^g(x, x)$.*

PROOF. By Proposition 3.1 and the last exercise, we have

$$
\begin{aligned}
e^{\lambda^2 g}(t, x, x) &= \sum_i e^{-(\lambda^{-2} \lambda_i) t} |\lambda^{-n/2} \varphi_i(x)|^2 \\
&= \lambda^{-n} \sum_i e^{-\lambda_i (\lambda^{-2} t)} |\varphi_i(x)|^2 \\
&= \lambda^{-n} e^g(\lambda^{-2} t, x, x).
\end{aligned}
$$

Thus

$$\frac{1}{(4\pi t)^{n/2}} \sum_k u_k^{\lambda^2 g}(x,x)t^k \;\sim\; e^{\lambda^2 g}(t,x,x) = \lambda^{-n}e^g(\lambda^{-2}t,x,x)$$

$$\sim\; \frac{\lambda^{-n}}{(4\pi(\lambda^{-2}t))^{n/2}} \sum_k u_k^g(x,x)(\lambda^{-2}t)^k$$

$$=\; \frac{1}{(4\pi t)^{n/2}} \sum_k \lambda^{-2k} u_k^g(x,x)t^k,$$

which implies

$$u_k^{\lambda^2 g} = \lambda^{-2k} u_k^g.$$

Fix i and n. Let $\mathcal{M} = \{m_j\}$ denote the set of monomials in P_i^n, and let $m_j(g)$ be this monomial computed for a metric g on an n-manifold. Each $m_j(g)$ is of the form $(\nabla^{k_1} R)^{p_1} \cdot \ldots \cdot (\nabla^{k_q} R)^{p_q}$ for some natural numbers $k_1, \ldots, k_q; p_1, \ldots, p_q$ depending on j (but not on g).

Lemma 3.28 *For any monomial* $(\nabla^{k_1} R)^{p_1} \cdot \ldots \cdot (\nabla^{k_q} R)^{p_q}$ *in* P_i^n, *we have*

$$i = \sum_{s=1}^q (1 + \frac{k_s}{2})p_s.$$

PROOF. We know that

$$u_i^{\lambda^2 g}(x,x) = \lambda^{-2i} u_i^g(x,x) = \lambda^{-2i} \sum_{m_j \in \mathcal{M}} m_j(g).$$

On the other hand, since $\nabla_{\lambda^2 g}^k R_{\lambda^2 g} = \lambda^{-2-k} \nabla_g^k R_g$ by Exercise 8, we have

$$u_i^{\lambda^2 g}(x,x) = \sum_{m_j \in \mathcal{M}} m_j(\lambda^2 g)$$

$$= \sum_{m_j \in \mathcal{M}} \lambda^{-\sum_s (2+k_s)p_s} m_j(g).$$

Combining these two equations gives two equal polynomials in λ, which implies that the coefficients and exponents in the polynomials must agree. This easily gives

$$-2i = -\sum_s (2+k_s)p_s$$

for each monomial.

In particular, when $i = 1$ each monomial in P_1^n satisfies $1 = \sum_s(1+(k_s/2))p_s$, which shows that the only possible monomial has $k_1 = 0$ and no other k's. In other words, P_1^n is a linear function with no constant term, and thus $u_1(x,x)$ is a linear function of the components of the curvature tensor at x, with no covariant derivative terms appearing. This monomial cannot be, say, some R_{ijkl} at x, as this component depends on the choice of coordinate chart, while $u_1(x, x)$ is independent of choice of coordinates (since $e(t, x, x)$ is). The expression $R_{ijkl}R^{ijkl}$ is independent of the coordinate chart, but is quadratic in the curvature. A little more experimenting should convince the reader that the only linear combination of curvature components that produces a well defined function $u_1(x, x)$ on a manifold is the scalar curvature $s(x) = R^{ij}_{ij}$. Thus there exists a constant C such that $u_1(x, x) = C \cdot s(x)$. We really only care whether or not the constant is nonzero.

Proposition 3.29 $u_1(x, x) = \frac{1}{6}s(x)$.

PROOF. We just sketch the proof; details are in [4, pp. 221–222]. Because P_1^n is a universal polynomial, it suffices to compute the constant C on one n-manifold, which we take to be S^n with the standard metric. On this manifold, we can calculate explicitly in Riemannian normal coordinates. For example, for y close to x, write $y = \exp_x rv$, for some $v \in T_x M$ with $|v| = 1$. Then $\det g(\exp_x rv) = (r^{-1} \sin r)^{n-1}$, and so

$$\phi(r) \equiv u_0(x, \exp_x rv) = \left(\frac{\sin r}{r}\right)^{-\frac{(n-1)}{2}}$$

$$= 1 + \frac{(n-1)}{12}r^2 + \frac{(n-1)(5n-1)}{1440}r^4 + O(r^6),$$

as $r \to 0$. Moreover, Theorem 2.63 gives, for $\tau \in [0, 1]$,

$$\Delta_y u_0(x, \exp_x \tau rv)$$
$$= \phi''(\tau r) + (n-1)\frac{\cos \tau r}{\sin \tau r}\phi'(\tau r)$$
$$= \left(\frac{n-1}{6} + O((\tau r)^2)\right)$$
$$+ (n-1)\left(\frac{1}{\tau r}(1 - \frac{(\tau r)^2}{3} + O((\tau r)^4))\right)\left(\frac{n-1}{6}\tau r + O(\tau r)^3\right).$$

We leave to the reader the computations of the Taylor expansions at $r = 0$ for $(r^{-1} \sin r)^{-(n-1)/2}$, etc. Performing the change of variables $\tau r = s$ in (3.12), and plugging all these expansions into the formula

$$u_1(x, \exp_x rv) = \frac{1}{\sqrt{\det \exp_x rv}} \int_0^1 \sqrt{\det \exp_x(\tau rv)}\Delta_y u_0(x, \exp_x \tau rv) \, d\tau,$$

we get $u_1(x, \exp_x ry) = \frac{n(n-1)}{6} + O(r^2)$. Thus

$$u_1(x, x) = \frac{n(n-1)}{6}$$

on S^n. By Exercise 8, Chapter 2, $s(x) = n(n-1)$ for all $x \in S^n$, which gives the result.

As anticipated, the calculation of u_1 gives another topological obstruction to manifolds having the same spectrum.

Theorem 3.30 *Let (M, g) and (N, h) be compact isospectral surfaces. Then M and N are diffeomorphic.*

PROOF. As noted in the proof of Corollary 3.25, we have

$$\int_M u_1^M(x, x) \, \mathrm{dvol}(x) = \int_N u_1^N(x, x) \, \mathrm{dvol}(x).$$

On a surface, the scalar curvature is twice the Gaussian curvature, so by the Gauss-Bonnet theorem,

$$6\pi\chi(M) = \int_M u_1^M(x, x) \, \mathrm{dvol}(x) = \int_N u_1^N(x, x) \, \mathrm{dvol}(x) = 6\pi\chi(N).$$

However, oriented surfaces with the same Euler characteristic are diffeomorphic.

The corresponding results for $u_2(x, x)$ are not encouraging. It is shown in [30] that

$$u_2(x, x) = \frac{1}{360}(2R_{ijkl}R^{ijkl}(x) + 2R_{jk}R^{jk}(x) + 5s^2(x) - 12\Delta s(x));$$

as with u_1, it is relatively easy to determine which curvature terms occur in u_2, but nontrivial to determine the coefficients of the terms. Unfortunately, $\int_M u_2$ has no topological significance. The best one can do is to prove results such as: if (M, g) and (N, h) are isospectral three-manifolds, and if M has constant sectional curvature, then so does N [4, Prop. E. IV.18].

The results for the higher u_i are downright discouraging. We have

$$u_3(x, x) = C \cdot R_{ijkl;m}R^{ijkl;m} + \dots,$$

where $R^{ijkl;m} = g^{mq}R^{ijkl}_{\quad ;q}$ and we have omitted approximately 40 terms. At the next stage, u_4 contains over 200 terms, some involving four covariant derivatives. Thus, while the higher u_i tell us more and more information about the geometry of isospectral manifolds, this information rapidly becomes too complicated to be of use.

Another possibility is to compute the asymptotics for the heat kernels on p-forms. As for functions, it can be shown that

$$e^p(t, x, y) = \sum e^{-\lambda_i t}\phi_i(x) \otimes \phi_i(y),$$

where $\{\phi_i\}$ is an orthonormal basis of $L^2\Lambda^k T^*M$ with $\Delta^k\phi_i = \lambda_i\phi_i$. As before,

$$\sum_i e^{-\lambda_i t} = \mathrm{Tr}(e^{-t\Delta^p}) = \int_M \mathrm{tr}\, e^p(t, x, x) \, dx,$$

108   CHAPTER 3. THE CONSTRUCTION OF THE HEAT KERNEL

where tr = tr_x denotes the pointwise trace on $\Lambda^k T_x^* M \otimes \Lambda^k T_x^* M$ given by the metric, as in Exercise 7.

Exercise 9: Let $I = (i_1, \ldots, i_p)$ and set $dx^I = dx^{i_1} \wedge \ldots \wedge dx^{i_p}$ as usual. Show that if $\alpha = \alpha_{IJ}(x,y)dx^I \otimes dy^J$ (where J is another multi-index of length p) is the local expression for a double form α, then $\mathrm{tr}_x \alpha(x,x) \, \mathrm{dvol}(x) = \alpha_{IJ} dx^I \wedge *dx^J$.

Patodi's construction of the heat kernel for forms yields an asymptotic expansion of double forms

$$e^p(t,x,x) \sim \frac{1}{(4\pi t)^{n/2}} \sum_{k=0}^{\infty} u_k^p(x,x)t^k.$$

Thus

$$\sum e^{-\lambda_i t} \sim \frac{1}{(4\pi t)^{n/2}} \sum_k \left(\int_M \mathrm{tr}\, u_k^p \right) t^k,$$

which as before implies that manifolds whose Laplacians on p-forms are isospectral have identical values for $\int \mathrm{tr}(u_k^p)$ for all k. However, scaling arguments as for functions show that $\mathrm{tr}(u_k^p)$ contains the same curvature terms as u_k^0, with possibly different coefficients. Thus there is no new topological information in these asymptotic expansions, and the new geometric information is not helpful.

Nevertheless, in the next chapter we will show that certain combinations of these asymptotic coefficients tr u_k^p for different p contain very important geometric and topological information.

3.4 Positivity of the Heat Kernel

In the Brownian motion model of heat flow on functions, we think of $e(t,x,y)$ as the probability that a particle starting at x will end up at y in time t. This is not a precise interpretation, but it suggests that $e(t,x,y)$ should be positive for all t,x,y, as we will show in this section.

In the proof, we only show $e(t,x,y) \geq 0$ and then appeal to a maximum principle to get strict inequality. We won't cover the maximum principle here, as we only need $e(t,x,y) \geq 0$ for the proof of the Chern-Gauss-Bonnet theorem in Chapter 4.

Theorem 3.31 For all t,x,y, we have $e(t,x,y) > 0$.

PROOF. We follow [16]. Pick a large integer N and $\epsilon > 0$ such that $u_k(x,y)$ is defined for $k = 1, \ldots, N$ for all x,y with $r(x,y) < \epsilon$. Fix $y_0 \in M$, $T > 0$, pick a small positive number δ and set $R = R_{\delta,T} = M \times [0,T] \setminus (B_\epsilon(y_0) \times [0,\delta])$.

As a function of x, we claim that $e(t,x,y_0), d_x e(t,x,y_0), \Delta_x e(t,x,y_0)$ are in $L^2(R)$. For $e(t,x,y_0)$, the only problem is the behavior of the heat kernel at $t = 0$. However, we have deleted the neighborhood of y_0 where the heat kernel blows up as $t \to 0$, and away from this neighborhood the heat kernel vanishes as $t \to 0$ by Exercise 5. Moreover, since $u_0(x, x) = 1 > 0$, there is a constant $K_N > 0$ such that $e(t, x, y) > -K_N t^N$ on R. Thus we can use Fubini's theorem to establish $e(t, x, y) \in L^2(R)$. The arguments for the other functions are similar; one must establish asymptotic expansions for these functions as in Exercise 5 via the (unproven) estimates at the end of Lemma 3.18.

We group some technical properties needed for the proof as an exercise.

Exercise 10: *(i) Let $u \in H_1(M)$. Set*

$$u_-(x) = \begin{cases} u(x), & u(x) \leq 0, \\ 0, & u(x) > 0. \end{cases}$$

Show that $u_- \in H_1(M)$. Hint: Take smooth functions u_i with $u_i \to u$ in H_1. Show by smoothing $(u_i)_-$ that $(u_i)_- \in H_1$. Now show $(u_i)_- \to u_-$ in H_1. Note that this proof does not work in H_2.

(ii) Consider d as a bounded operator taking $H_1(M)$ to $L^2 \Lambda^1 T^* M$. Show that for $u \in H_1$

$$du_-(x) = \begin{cases} du(x), & u(x) \leq 0, \\ 0, & u(x) > 0. \end{cases}$$

(iii) For $f \in L^2 \Lambda^1 T^* M, u \in H_1(M)$, show that $\langle f, du \rangle = \langle \delta f, u \rangle$, where $\langle \, , \, \rangle$ denotes the L^2 inner products.

(iv) Show that $\partial_t |e(t, x, y)_-|^2 = 2 \langle \partial_t e(t, x, y)_-, e(t, x, y)_- \rangle$. Hint: derive an expression for $\partial_t e(t, x, y)_-$ similar to the expression for du_- in (ii).

(v) Let M be a manifold with boundary. Use Stokes' theorem to show that for smooth functions f, g on M,

$$\int_M \langle df, dg \rangle \mathrm{dvol} = \int_M \langle \Delta f, g \rangle \mathrm{dvol} - \int_{\partial M} *f dg.$$

Conclude that the same equation holds if $g \in H_1(M)$.

Set $e = e(t, x, y_0)$. Using (ii), (iii) and (v) (Stokes' theorem for each value of $t \in [0, \delta]$), we get

$$0 \leq \int_R \langle de, de_- \rangle \, \mathrm{dvol}(x) dt$$

$$= \int_R \langle \Delta e, e_- \rangle - \int_{\partial B_\epsilon(y_0) \times [0,\delta]} *e_- de$$

$$= - \int_R \langle \partial_t e, e_- \rangle - \int_{\partial B_\epsilon(y_0) \times [0,\delta]} *e_- de. \tag{3.32}$$

Since $\partial_t e_- = \partial_t e$ whenever $e_- \neq 0$, the first term in the last line of (3.32) is

$$- \int_R \langle \partial_t e, e_- \rangle = - \int_R \langle \partial_t e_-, e_- \rangle = -\frac{1}{2} \int_R \partial_t |e_-|^2$$

$$= -\frac{1}{2} \int_{M \times \{T\}} |e_-|^2 + \frac{1}{2} \int_{B_\epsilon(y_0) \times \{\delta\}} |e_-|^2. \tag{3.33}$$

The additional integral in the last line over $(M \setminus B_\epsilon(y_0)) \times \{0\}$ vanishes by Exercise 3. It follows from the asymptotic expansion in Exercise 5 that the last terms in (3.32), (3.33) vanish as $\delta \to 0$. (For the term in (3.33), we use again that $u_0(x, y_0) > 0$ for ϵ small enough, and so $e_- = 0$ on this neighborhood.) Thus we obtain

$$0 \le -\frac{1}{2} \int_{M \times \{T\}} |e_-|^2,$$

and so $e_- \equiv 0$. Now that we have $e \ge 0$, the maximum principle for parabolic operators implies $e > 0$ [57].

Chapter 4

The Heat Equation Approach to the Atiyah-Singer Index Theorem

The Atiyah-Singer index theorem is a deep generalization of the classical Gauss-Bonnet theorem, including as special cases the Chern-Gauss-Bonnet theorem, the Hirzebruch signature theorem, and the Hirzebruch-Riemann-Roch theorem. Although the index theorem is about 35 years old at this point, it continues to have new applications in areas as apparently diverse as number theory and mathematical physics. The index theorem and its various generalizations (families index theorem, K-theoretic versions, etc.) admit many interpretations. We will choose the point of view that the index theorem expresses topological quantities in terms of geometric ones, just as in the Gauss-Bonnet theorem. This viewpoint leads to a heat equation proof of the index theorem, suggested by McKean and Singer [43] in the late 1960s and established by Gilkey [29], Patodi [55, 56], and Atiyah, Bott and Patodi [1] in the early 1970s. The heat equation method has since been refined by Getzler [28] (cf. [5]).

In this chapter, we will give a complete heat equation proof for the Chern-Gauss-Bonnet theorem, and state without proof the Hirzebruch signature theorem, the Hirzebruch-Riemann-Roch theorem, and the Atiyah-Singer index theorem. Complete proofs can be found in [5] and [30]. We have also included a short introduction to characteristic classes.

4.1   The Chern-Gauss-Bonnet Theorem

The key ideas in the heat equation method are (i) by Chapter 1, the long time behavior of the heat operator for the Laplacian on forms is controlled by the topology of the manifold in the form of the de Rham cohomology, (ii) the short time behavior is controlled by the geometry of the asymptotic expansion, as explained in Chapter 3, and (iii) certain combinations of heat operators will have time independent behavior. This combination of topological information therefore has a geometric interpretation, made explicit by the Chern-Gauss-Bonnet theorem.

4.1.1   The Heat Equation Approach

Let M be a closed, oriented, connected n-manifold. We begin with a crucial observation of McKean and Singer [43]:

Lemma 4.1 *For $\lambda \in \mathbf{R}^+$, let E_λ^q be the (possibly trivial) λ-eigenspace for Δ^q. Then the sequence*

$$0 \longrightarrow E_\lambda^0 \xrightarrow{\ d\ } \ldots \xrightarrow{\ d\ } E_\lambda^n \longrightarrow 0$$

is exact.

PROOF. If $\omega \in E_\lambda^q$, then $\Delta^{q+1} d\omega = d\Delta^q \omega = \lambda d\omega$, so $d\omega \in E_\lambda^{q+1}$. Thus the sequence is well defined, and of course has $d^2 = 0$. If $\omega \in E_\lambda^q$ has $d\omega = 0$, then $\lambda\omega = \Delta^q\omega = (\delta d + d\delta)\omega = d\delta\omega$, and so $\omega = d\left(\frac{1}{\lambda}\delta\omega\right)$, since $\lambda \neq 0$.

Exercise 1: *Show that the lemma is equivalent to the statement that*

$$d + \delta : \bigoplus_k E_\lambda^{2k} \longrightarrow \bigoplus_k E_\lambda^{2k+1}$$

is an isomorphism. Conclude that

$$\sum_q (-1)^q \dim E_\lambda^q = 0.$$

Corollary 4.2 *Let $\{\lambda_i^q\}$ be the spectrum of Δ^q. Then*

$$\sum_q (-1)^q \sum_i e^{-\lambda_i^q t} = \sum_q (-1)^q \dim \mathrm{Ker}\ \Delta^q.$$

PROOF. By the exercise,

$$\sum_q (-1)^q \sum_i e^{-\lambda_i^q t} = \sum_q (-1)^q {\sum_i}' e^{-\lambda_i^q t},$$

where the sum ${\sum}'$ is only over those i for which $\lambda_i^q = 0$. Thus

$${\sum}' e^{-\lambda_i^q t} = \dim \mathrm{Ker}\ \Delta^q.$$

As a result,

$$\sum_q (-1)^q \mathrm{Tr}\, e^{-t\Delta^q} = \sum_q (-1)^q \sum_i e^{-\lambda_i^q t}$$

is independent of t, which of course means that its long time behavior is the same as its short time behavior. We know that the long time behavior of $\mathrm{Tr}\, e^{-t\Delta^q}$ is given by the de Rham cohomology, while the short time behavior is controlled by the geometry of the manifold. In particular,

$$
\begin{aligned}
\chi(M) &= \sum_q (-1)^q \dim H_{dR}^q(M) = \sum_q (-1)^q \dim \mathrm{Ker}\, \Delta^q \\
&= \sum_q (-1)^q \mathrm{Tr}\, e^{-t\Delta^q} \\
&= \sum_q (-1)^q \int_M \mathrm{tr}\, e^q(t, x, x)\, \mathrm{dvol} ,
\end{aligned}
\tag{4.3}
$$

which yields

$$\chi(M) \sim \frac{1}{(4\pi t)^{\frac{n}{2}}} \sum_{k=0}^{\infty} \left(\int_M \sum_{q=0}^{n} (-1)^q \mathrm{tr}\, u_k^q(x, x)\, \mathrm{dvol} \right) t^k,$$

where the u_k^q are the coefficients in the asymptotic expansion of $\mathrm{Tr}\, e^{-t\Delta^q}$. Since $\chi(M)$ is independent of t, only the constant term on the right hand side can be nonzero.

Theorem 4.4

$$(4\pi)^{-n/2} \int_M \sum_{q=0}^{n} (-1)^q \mathrm{tr}\, u_k^q(x, x)\, \mathrm{dvol} = \begin{cases} 0, & k \neq \frac{n}{2}, \\ \chi(M), & k = \frac{n}{2}, \ \dim M \text{ even.} \end{cases}$$

Exercise 2: *Re-prove the following result from Chapter 1: if $\pi : M' \to M$ is an ℓ-fold cover, then $\chi(M') = \ell \cdot \chi(M)$. Hint: We may assume that M is even dimensional. Fix a metric g on M and give M' the pullback metric. Show that $\mathrm{tr}\, u_k^{q,M'}(x', x') = \mathrm{tr}\, u_k^{q,M}(\pi(x'), \pi(x'))$, for all $x' \in M'$ and all q, k.*

Exercise 3: *Show that $\mathrm{tr}\, u_k^q(x, x) = \mathrm{tr}\, u_k^{n-q}(x, x)$. Hint: first show that $*e^{-t\Delta^q} = e^{-t\Delta^{n-q}} *$.*

Corollary 4.5 (Gauss-Bonnet Theorem) *Let M be a closed oriented surface with Gaussian curvature K and area element dA. Then*

$$\chi(M) = \frac{1}{2\pi} \int_M K\, dA.$$

PROOF. By the last theorem and exercise, together with Proposition 3.29,

$$\chi(M) = \frac{1}{4\pi} \int_M \sum_{q=0}^{2} (-1)^q \mathrm{tr}\ u_1^q\ dA = \frac{1}{4\pi} \int_M (\frac{2K}{3} - \mathrm{tr}\ u_1^1)\ dA. \qquad (4.6)$$

We know that $\mathrm{tr}\ u_1^1(x, x) = C \cdot s(x) = 2C \cdot K(x)$ for some constant C. Since the standard S^2 has constant Gaussian curvature one, we get

$$2 = \frac{1}{2\pi} \int_{S^2} (\frac{1}{3} - C)\ dA = \frac{1}{2\pi}(\frac{1}{3} - C)(4\pi).$$

Therefore, $C = -(2/3)$, and (4.6) becomes $\chi(M) = (2\pi)^{-1} \int_M K\ dA$.

Of course, this is a very elaborate and ungeometric proof of the Gauss-Bonnet theorem, but it shows that Gauss-Bonnet generalizes to a result holding in all dimensions. In odd dimensions, it only re-proves the result $\chi(M) = 0$ (check), but in even dimensions it shows that the Euler characteristic can be obtained by integrating an expression in the curvature and its covariant derivatives. On the other hand, we expect this expression to be quite unmanageable in high dimensions, as it involves an enormous number of curvature terms with many covariant derivatives.

In contrast, Chern's generalization of the Gauss-Bonnet theorem [17] from 1944 is as follows:

Theorem 4.7 (Chern-Gauss-Bonnet Theorem) *Let* dim $M = n$ *be even. Then*

$$\chi(M) = \int_M \omega,$$

where the Euler form ω is given in a local orthonormal frame by

$$\omega = c_n \sum_{\sigma, \tau \in \Sigma_n} (\mathrm{sgn}\ \sigma)(\mathrm{sgn}\ \tau) R_{\sigma(1)\sigma(2)\tau(1)\tau(2)} \cdots R_{\sigma(n-1)\sigma(n)\tau(n-1)\tau(n)}\ \mathrm{dvol}\ ,$$

with Σ_n the group of permutations of $\{1, \ldots, n\}$ and

$$c_n = \frac{(-1)^{n/2}}{(8\pi)^{n/2}(n/2)!}.$$

Computing in an orthonormal frame can be done as in (2.34). Equivalently, we can choose a coordinate chart $(x^1, \ldots, x^n)$ and an orthonormal frame $(X_1, \ldots, X_n)$ of TM near a given point, and transform the $R_{ijk\ell}$ for the x coordinates into the corresponding $R_{ijk\ell} = R_{ijk\ell}^{\mathrm{fr}}$ for the frame:

$$R_{ijk\ell}^{\mathrm{fr}} = R_{pqrs} X_i(x^p) X_j(x^q) X_k(x^r) X_\ell(x^s).$$

The $(-1)^{n/2}$ factor in c_n is due to our sign convention for $R_{ijk\ell}$ in (2.9). To make sure the signs are right, we check the case dim $M = 2$. By (2.19), the Euler form equals

$$-\frac{1}{8\pi}(R_{1212} - R_{1221} - R_{2112} + R_{2121})\mathrm{dvol} = -\frac{1}{2\pi} R_{1212}\ \mathrm{dvol} = \frac{1}{2\pi} K dA.$$

Thus the Chern-Gauss-Bonnet theorem generalizes the classical Gauss-Bonnet theorem.

In higher dimensions it is more difficult to extract the curvature information encoded in the Euler form. For example, if M has constant negative curvature, then by Hirzebruch proportionality [33] the sign of each term in the Euler form is $(-1)^{\dim M/2}$. The Chern-Gauss-Bonnet theorem thus confirms the Hopf conjecture for compact manifolds of constant negative curvature (regardless of the existence of such manifolds).

Moreover, we have the following topological obstruction to a manifold admiting a flat metric.

Corollary 4.8 *Let M be a compact oriented manifold admitting a flat metric. Then $\chi(M) = 0$.*

While this is not helpful in odd dimensions, it provides many examples of even dimensional manifolds (spheres, projective spaces) which do not admit flat metrics.

Exercise 4: *Show that the Euler form for a four-manifold equals*

$$\frac{1}{16\pi^2} \sum_q (-1)^q \mathrm{tr}\, u_2^q(x,x).$$

Warning: This is a nontrivial calculation. As in the proof of Gauss-Bonnet, you must calculate $\mathrm{tr}\, u_2^1(x,x) = AR_{ijkl}R^{ijkl} + BR_{ij}R^{ij} + Cs^2 + D\Delta s$ *by computing $\chi(M)$ on enough four-manifolds to determine the constants A, B, C, D. Then you must do the same for* $\mathrm{tr}\, u_2^2(x,x)$. *Finally, you have to rewrite the Euler form to see that it agrees with $\sum_q (-1)^q \mathrm{tr}\, u_2^q(x,x)$. In particular, the coefficient of Δs that appears in this sum must vanish, since Δs involves two covariant derivatives of the curvature (why?), and the Euler form has no covariant derivatives. For details, see [4], [30].*

These considerations led McKean and Singer to conjecture that in all even dimensions,

$$\omega(x) = \frac{1}{(4\pi)^{n/2}} \sum_q (-1)^q \mathrm{tr}\, u_{\frac{n}{2}}^q(x,x).$$

As in the last exercise, this conjecture would imply a "fantastic cancellation" of the numerous terms containing covariant derivatives of the curvature in the individual $\mathrm{tr}\, u_{n/2}^q(x,x)$. This conjecture was first shown by Patodi in 1972 using very technical classical tensor calculus, and his approach was extended in [1] to a heat equation proof of the Atiyah-Singer index theorem (see §4.2.3). A much clearer proof was given by Getzler around 1983, who realized that the necessary cancellations reflected properties of endomorphisms of Clifford algebras. In the next subsection, we will prove the Chern-Gauss-Bonnet theorem along these lines.

It might be wondered whether the Chern-Gauss-Bonnet theorem can be refined to show that the individual Betti numbers are given by the integral of a universal local expression $e^k(g) = e^k(g)(x)$ in the Riemannian metric. In fact, no such formula is possible. For if $\beta^k(M) = \int_M e^k(g)$, and $\pi : M' \to M$ is an ℓ-fold cover of M, then as in Exercise 2,

$$\beta^k(M') = \int_{M'} e^k(\pi^*g) = \ell \int_M e^k(g) = \ell\beta^k(M),$$

since $\pi : (M', \pi^*g) \to (M, g)$ is a local isometry. However, the Betti numbers of covering spaces do not satisfy this relation in general.

4.1.2 Proof of the Chern-Gauss-Bonnet Theorem

In this subsection we will present a fermion calculus proof of the Chern-Gauss-Bonnet theorem due to Parker [53]. Other fermion calculus proofs exist [18, 23], but involve either stochastic methods or more involved functional analysis.

We may assume that dim $M = n$ is even. To fix the notation, let $x, y \in M$ have distance $r = r(x, y)$ small enough so that there exists a unique minimal geodesic $\gamma(s)$ from x to y. We define an approximate heat kernel for the Laplacian on functions by

$$\bar{E}(t, x, y) = \frac{1}{(4\pi t)^{n/2}} e^{-r^2(x,y)/4t} u_0(x, y). \tag{4.9}$$

Recall from (3.11) that $u_0(x, y) = (\det(P(x, y)d\exp_x))^{-1/2}$, where $P(x, y) : T_y M \to T_x M$ denotes parallel translation along γ.

Parallel translation induces a linear isomorphism $P = P(x, y) : \Lambda^* T_y^* M \to \Lambda^* T_x^* M$. Let $\Delta = \nabla^*\nabla + R$ be the Weitzenböck decomposition of the Laplacian on forms of mixed degree. We define an approximate heat kernel on forms by

$$\bar{e}(t, x, y) = \bar{E}(t, x, y) \cdot \exp\left(-\int_0^t (P^{-1}RP)(\gamma(s), x)\, ds\right) P(x, y). \tag{4.10}$$

(On functions, $\bar{e} = \bar{E}$, and on n-forms we omit the exponential term.) More explicitly, let $\{\theta^I(s)\}$ be a parallel orthonormal frame of $\Lambda^* T_{\gamma(s)}^* M$ and set $R_{\gamma(s)}\theta^I(s) = a_J^I(s)\theta^J(s)$. For a form $v = v_I\theta^I$ defined along γ,

$$(P^{-1}RP(\gamma(s), x))(v) = v_I(x)a_J^I(s)\theta^J(x).$$

The motivation for this formula will be given below.

For fixed t, x, the heat kernel $e(t, x, x)$ for the Laplacian Δ acting on forms of mixed degree can be considered as an endomorphism of $\Lambda^* T_x^* M$. In the notation of §2.2.2, this endomorphism has a component indexed by $\{1, \ldots, n\}\{1, \ldots, n\}$, which we denote by $e(t, x, x)_{2n}$. For any endomorphism A of $\Lambda^* T_x^* M$, we let $\exp(A)$ denote the power series expansion $\sum(A^k/k!)$; in all the cases we will consider, A will be nilpotent, so this sum will be finite.

Theorem 4.11 *There exists a positive constant C such that for all $t \in (0, 1]$*

$$\|(e - \bar{e})(t, x, x)_{2n}\| \leq Ct. \tag{4.12}$$

Here the norm of an element in

$$\mathrm{Hom}(\Lambda^* T_y^* M, \Lambda^* T_x^* M) \cong (\Lambda^* T_y^* M)^* \otimes (\Lambda^* T_x^* M)$$

is induced by the Riemannian norms on $T_x M, T_y M$.

The proof of the theorem will be given below. As for its significance, we know that the parametrix H_k for the heat kernel satisfies $\|e(t, x, x) - H_k(t, x, x)\|$ $\leq Ct$ for $k \gg 0$, but H_k involves many covariant derivatives of the curvature. In contrast, $\bar{e}$ has no curvature derivatives, and the theorem says that the supertrace of $e - \bar{e}$ is already $O(t)$.

In particular, we see from (4.3) that

$$\begin{aligned}
\chi(M) &= \int_M \mathrm{tr}((-1)^F e(t, x, x)) \, \mathrm{dvol} \\
&= \int_M \mathrm{tr}((-1)^F e(t, x, x)_{2n}) \, \mathrm{dvol} ,
\end{aligned}$$

by Proposition 2.30. Assuming the theorem, we obtain

$$\chi(M) = \int_M \mathrm{tr}((-1)^F \bar{e}(t, x, x)_{2n}) \mathrm{dvol} + O(t)$$

as $t \downarrow 0$. $\chi(M)$ is independent of t, so

$$\chi(M) = \int_M \mathrm{tr}((-1)^F \bar{e}(t, x, x)_{2n}) \, \mathrm{dvol} .$$

Since $\bar{E}(t, x, x) = (4\pi t)^{-n/2}$ and $P(x, x) = \mathrm{Id}$, we find

$$\chi(M) = \frac{1}{(4\pi t)^{n/2}} \int_M \mathrm{tr}((-1)^F (\exp(-tR(x)))_{2n}) \, \mathrm{dvol} . \tag{4.13}$$

We claim that

$$\exp(-tR)_{2n} = \frac{1}{k!}(-1)^k t^k R^k + O(t^{k+1}), \tag{4.14}$$

where $k = n/2$. Writing $R = R_{ijk\ell} a_i^* a_j a_k^* a_\ell$ and expanding $\exp(-tR)$ as a power series, we see that the first term in $\exp(-tR)_{2n}$ is $(1/k!)(-1)^k t^k R^k$ since no coefficient of a lower power of t contains n a's and n a^*'s. This may not be the only term contributing to $(e^{-tR})_{2n}$; higher powers of $-tR$ may have components of degree $2n$, since reshuffling the a's and a^*'s to get them in increasing order introduces terms of lower degree via (2.27). However, these higher contributions are all $O(t^{k+1})$, so the claim is established.

By (4.13), (4.14), we have

$$\chi(M) = \frac{(-1)^k}{(4\pi)^k k!} \int_M \operatorname{tr}((-1)^F R^k) \, \text{dvol} \; + \mathrm{O}(t),$$

and as above we conclude that

$$\chi(M) = \frac{(-1)^k}{(4\pi)^k k!} \int_M \operatorname{tr}((-1)^F R^k) \, \text{dvol} \, .$$

It follows from Lemma 2.35 that

$$\frac{(-1)^k}{(4\pi)^k k!} \operatorname{tr}((-1)^F R^k) = \omega,$$

in the notation of Theorem 4.7. This completes the proof of the Chern-Gauss-Bonnet theorem, modulo the proof of Theorem 4.11.

The proof of this latter theorem proceeds in a number of steps.

Lemma 4.15 *There exists a constant $C > 0$ such that*

$$\|(e - \bar{e})(t, x, y)\| \le Ct\|\bar{e}(t, x, y)\| \tag{4.16}$$

for $t \le 1$ and x, y close.

PROOF. In the case of functions, we have

$$
\begin{aligned}
|(e - \bar{E})(t, x, y)| &= \left| \frac{e^{-r^2(x,y)/4t}}{(4\pi t)^{n/2}} \left(\sum_{k=0}^{N} u_k(x, y)t^k + \mathrm{O}(t^N) - u_0(x, y) \right) \right| \\
&= \left| \frac{e^{-r^2(x,y)/4t}}{(4\pi t)^{n/2}} u_0(x, y) \left(\sum_{k=1}^{N} \frac{u_k(x, y)}{u_0(x, y)} t^k + \mathrm{O}(t^N) \right) \right| \\
&\le Ct|\bar{E}(t, x, y)|,
\end{aligned}
$$

since $u_0(x, y)$ is nonzero for x, y close.

In the case of forms, we first set

$$e_1(t, x, y) = \bar{E}(t, x, y)P(x, y).$$

This is just the first parametrix for the heat kernel on forms. As above, we have

$$\|(e - e_1)(t, x, y)\| \le Ct\|e_1(t, x, y)\|.$$

Moreover,

$$\|(\bar{e} - e_1)(t, x, y)\|$$

$$= \left\| \bar{E}(t,x,y) \cdot \left[\left(\exp(- \int_0^t (P^{-1}RP)(\gamma(s),x)\, ds) \right) P(x,y) - P(x,y) \right] \right\|$$

$$= \left\| \bar{E}(t,x,y) \cdot [(\mathrm{Id} - \int_0^t P^{-1}RP + \frac{1}{2}(\int_0^t P^{-1}RP)^2 + \ldots)P(x,y) \right.$$

$$\left. -P(x,y)] \right\|$$

$$= \left\| \bar{E}(t,x,y) \cdot [(\mathrm{Id} - R(x)t + \mathrm{O}(t^2))P(x,y) - P(x,y)] \right\|$$

$$\leq \quad Ct\|e_1(t,x,y)\|. \tag{4.17}$$

(In the second line, we used the definition of the exponential, and in the fourth line we applied Taylor's theorem to the integrals.) Combining the last two equations gives

$$\|(e - \bar{e})(t,x,y)\| \leq Ct\|e_1(t,x,y)\|. \tag{4.18}$$

We have to replace e_1 on the right hand side of (4.18) with $\bar{e}$. From (4.17) we have $\|te_1(t,x,y) - t\bar{e}(t,x,y)\| \leq Ct^2 e(t,x,y)$, and substituting this in the right hand side of (4.18) gives $\|(e - \bar{e})(t,x,y)\| \leq Ct\|\bar{e}(t,x,y)\| + C^2 t^2\|e(t,x,y)\|$. Continuing this process leads to

$$\|(e - \bar{e})(t,x,y)\| \quad \leq \quad Ct\|\bar{e}(t,x,y)\| + C^2 t^2\|\bar{e}(t,x,y)\| + \ldots + C^N t^N\|\bar{e}(t,x,y)\|$$
$$+C^N t^N\|e(t,x,y)\|,$$

for any $N > 0$. The last term is $\mathrm{O}(t^2)$ for N sufficiently large, since $\|e(t,x,y)\| = \mathrm{O}(t^{-n/2})$, and so we obtain

$$\|(e - \bar{e})(t,x,y)\| \leq Ct\|\bar{e}(t,x,y)\|,$$

which finishes the proof.

Remark: Using the techniques in the last paragraph, we can also show that for the heat kernel on functions,

$$|(e - \bar{E})(t,x,y)| \leq Ct|e(t,x,y)|. \tag{4.19}$$

We now fix $x \in M$ and set $r = r(y) = r(x,y)$ to be the distance from x to y. There exists $\rho > 0$ such that there is a unique minimal geodesic from x to y whenever $r(y) < \rho$. Choose a smooth bump function $\phi(y) = \phi(r(y))$ with $\phi(y) \in [0,1]$, $\phi(r) = 1$ for $r < \rho/4$, and $\phi(r) = 0$ for $r > \rho/2$. Since $\bar{e}$ is a parametrix, $\phi\bar{e}(t,x,y)$ approaches the delta function centered at x as $t \to 0$. The defining properties (1.24) of the heat kernel give

$$\left. \begin{aligned} (\partial_t + \Delta_x)(e - \phi\bar{e})(t,x,y) &= -(\partial_t + \Delta_x)\phi\bar{e}(t,x,y), \\ \lim_{t \to 0}(e - \phi\bar{e})(t,x,y) &= 0, \end{aligned} \right\} \tag{4.20}$$

where the last line is an equality in $L^2(\Lambda^* \otimes \Lambda^*)$.

Exercise 5: *Given a time dependent form* $\psi(x,t) \in \Lambda^* T^* M \otimes \mathbf{R}$, *show that the solution* $\phi(x,t)$ *to* $(\partial_t + \Delta)\phi(x,t) = \psi(x,t)$ *is given by*

$$\phi(x,t) = \int_0^t ds \int_M e(t-s,x,y) \wedge *_y \psi(y,s),$$

where $*_y$ *indicates that the star operator acts in the y variable.*

By this exercise, the solution to (4.20) is

$$(e - \phi\bar{e})(t,x,y) = -\int_0^t ds \int_M e(t-s,x,z) \wedge *_z (\partial_t + \Delta)\phi\bar{e}(s,z,y). \qquad (4.21)$$

To proceed with this Duhamel-type formula, we need to calculate

$$(\partial_t + \Delta)\phi\bar{e}(s,z,y).$$

As a preliminary step, we compute $(\partial_t + \Delta)\bar{e}(t,x,y)$ for x,y close. The computation is fairly horrendous, so keep in mind that we really only care about the short time behavior of these terms.

Set

$$B(t,x,y) = \int_0^t P^{-1} R P(\gamma(s),x) \, ds,$$

so that $\bar{e} = \bar{E}e^{-B}P$. We have $(\partial_t + \Delta)\bar{E} = u_0^{-1}\Delta u_0 \bar{E}$ from (3.13). Combining this with Exercise 4, Chapter 3, we get

$$
\begin{aligned}
(\partial_t + \Delta)\bar{e}(t,x,y) &= (\partial_t + \Delta)\bar{E}e^{-B}P + \bar{E}\partial_t(e^{-B}P) + \bar{E}\Delta(e^{-B}P) \\
&\quad -2\langle \nabla(e^{-B}P), \nabla\bar{E}\rangle \\
&= u_0^{-1}\Delta u_0\bar{e} - \bar{E}e^{-B}P_t^{-1}R_t P_t P + \bar{E}\Delta(e^{-B}P) \\
&\quad -2\langle \nabla(e^{-B}P), \nabla\bar{E}\rangle,
\end{aligned}
$$

where $P_t = P(\gamma(t),x), R_t = R(\gamma(t))$. Here we use $\nabla\bar{E}$ and $d\bar{E}$ interchangeably.

Exercise 6: *(i) For fixed* x,y, *parallel translation* P *is an element of* $\mathrm{Hom}(T_y M, T_x M)$. *The bundle* $\mathrm{Hom}(TM,TM) \cong TM \otimes T^* M$ *has a natural connection* $\bar{\nabla}$ *induced from the Levi-Civita connections* ∇, ∇' *on* $TM, T^* M$, *respectively, namely*

$$\bar{\nabla}(\omega \otimes \eta) = \nabla\omega \otimes \eta + \omega \otimes \nabla'\eta$$

(see Exercise 11, Chapter 2). Check that this connection corresponds to a connection on $\mathrm{Hom}(TM,TM)$, *denoted* ∇ *for short, characterized by*

$$\nabla_X(A(V)) = (\nabla_X A)(V) + A(\nabla_X V),$$

for $A \in \Gamma(\mathrm{Hom}(TM,TM))$, $V \in \Gamma(TM)$, $X \in T_x M$.

(ii) Show that parallel translation is parallel in radial directions. In particular, $\nabla_X P = 0$ for all $X \in T_x M$. Hint: extend V to a parallel vector field along radial geodesics at x, and let X be the vector field ∂_r near x. $\nabla_r(P(V)) = 0$ by definition of parallel translation, and similarly $\nabla_r V = 0$. Thus $\nabla_r P = 0$.

From the exercise, $\nabla(e^{-B}P) = (\nabla e^{-B})P$, so we may drop the parentheses. Note that $P_t^{-1}R_{\gamma(t)}P_t$ is differentiable in t. Recalling that $P = P(y, x)$, $R_0 = R_x$, we have

$$
\begin{aligned}
(\partial_t + \Delta)\bar{e}(t, x, y) &= u_0^{-1}\Delta u_0 \bar{e} + \bar{E}e^{-B}\frac{(-P_t^{-1}R_t P_t + P_0^{-1}R_0 P_0)}{t}tP \\
&\quad -\bar{E}e^{-B}R_0 P + \bar{E}(\nabla^*\nabla + R_0)(e^{-B}P) \\
&\quad -2\langle \nabla e^{-B}P, \nabla \bar{E}\rangle \\
&= u_0^{-1}\Delta u_0 \bar{e} + \bar{E}e^{-B}O(t)P + \bar{E}\nabla^*\nabla(e^{-B}P) \\
&\quad -2\langle \nabla e^{-B}P, \nabla \bar{E}\rangle \\
&= u_0^{-1}\Delta u_0 \bar{e} + \bar{E}e^{-B}O(t)P + \nabla^*\nabla(e^{-B}P)P^{-1}e^B \bar{e} \\
&\quad -2\langle \nabla e^{-B}P, \nabla u_0\rangle u_0^{-1}P^{-1}e^B \bar{e} \\
&\quad -\frac{1}{2t}\langle \nabla e^{-B}P, \nabla r^2\rangle P^{-1}e^B \bar{e}.
\end{aligned}
$$

In summary, for x, y close we have

$$(\partial_t + \Delta)\bar{e}(t, x, y) = G\bar{e},$$

for

$$
\begin{aligned}
G(t, x, y) &= O(t) + u_0^{-1}\Delta u_0 + \nabla^*\nabla(e^{-B}P)P^{-1}e^B \\
&\quad -2\langle \nabla e^{-B}P, \nabla u_0\rangle u_0^{-1}P^{-1}e^B \\
&\quad -\frac{1}{2t}\langle \nabla e^{-B}P, \nabla r^2\rangle P^{-1}e^B.
\end{aligned}
$$

Note the $O(1/t)$ term.

Exercise 7: *Show that the $O(t)$ term is an endomorphism which has an expansion*

$$A_1 t + A_2 t^2 + \ldots + A_N t^N + O(t^{N+1})$$

for any N, where each A_k contains two a's and two a^*'s. Hint: The Taylor series expansion for B in t gives the A_k. In particular, A_1 is R_0. Then $A_2 = \nabla_r R_0$. Since $\nabla_r \theta^i = 0$ at x, we have $\nabla_r a_i = \nabla_r a_i^* = 0$ at x. Thus $A_2 = -\partial_r R_{ijk\ell}a_i^* a_j a_k^* a_\ell$. For the higher A_k, note that $\nabla_r^{k-1}a_i^* = b_j^i a_j^*$, where $\nabla_r^{k-1}\theta^i = b_j^i \theta^j$.

Since ϕ is a radial function, by Gauss' lemma we get

$$
\begin{aligned}
(\partial_t + \Delta_x)\phi\bar{e} &= \phi(\partial_t + \Delta_x)\bar{e} + \Delta\phi \cdot \bar{e} - 2\langle \nabla\phi, \nabla\bar{e}\rangle \\
&= \phi G\bar{e} + \Delta\phi \cdot \bar{e} - 2\partial_r\phi \cdot \nabla_r \bar{e}.
\end{aligned}
$$

Now

$$
\begin{aligned}
\nabla_r \bar{e} &= \nabla_r (\bar{E} e^{-B}) P \\
&= -\frac{r}{2t} \bar{e} + u_0 (\nabla_r u_0) \bar{e} + \bar{E} (\nabla_r e^{-B}) P \\
&= -\frac{r}{2t} \bar{e} + u_0 (\nabla_r u_0) \bar{e} + (\nabla_r e^{-B}) e^B \bar{e}.
\end{aligned}
$$

This gives

$$
\begin{aligned}
(\partial_t + \Delta_x) \phi \bar{e} &= \phi G \bar{e} + \Delta \phi \cdot \bar{e} \\
&\quad -2 \partial_r \phi (-\frac{r}{2t} \bar{e} + u_0 \nabla_r u_0 \bar{e} - \bar{e} \nabla_r B).
\end{aligned}
$$

Thus we can write

$$
(\partial_t + \Delta_x) \phi \bar{e} = H(t, x, y) \bar{e}, \tag{4.22}
$$

where H blows up at worst like $1/t$ as $t \to 0$. Because of the presence of ϕ, this equation holds for all $x, y \in M$.

We wish to substitute (4.22) back into (4.21) in order to estimate $e - \bar{e}$. However, the $H \bar{e}$ term may be unintegrable near $t = 0$ due to the t^{-1} term on the region where ϕ is nonconstant. The following technical lemma shows that no such problem occurs.

Lemma 4.23 Set $\mu = \rho/12$ and let $\Omega = B_{\rho/2}(x) \backslash B_{\rho/4}(x)$. There exist constants $C, \delta > 0$ such that for $0 < t \leq 1$ and $r = r(x, y) < \mu$,

$$
\left\| \int_0^t ds \int_\Omega e(t - s, x, z) \wedge *_z H \bar{e}(s, z, y) \right\| \leq C \exp(-\delta/t) \bar{E}(t, x, y). \tag{4.24}
$$

PROOF: We use the convention that C denotes a constant whose value may change from line to line in the proof. As in Exercise 3, Chapter 3, the compactness of $\{(t, x, z) \in [0, 1] \times M \times M : r(x, z) \geq 3\mu\}$ implies that the smooth function $\|e(t, x, z)\|$ is bounded on this set. Similarly, there exist constants c_1, c_2 such that $|H| \leq c_1/t$ and $c_2^{-1} \leq \|u_0 e^{-B} P\| \leq c_2$ on $\{(s, z, y) \in [0, 1] \times M \times M : r(z, y) \leq \rho/2\}$. Thus the left hand side of (4.24) is bounded above by

$$
C \int_0^t \frac{ds}{s} \int_\Omega (4\pi s)^{-n/2} \exp(-r^2(z, y)/4s) \, \mathrm{dvol}_z .
$$

Since $r(z, y) \geq r(z, x) - r(x, y) \geq 2\mu \geq 2r + 2\delta$ for some $\delta > 0$, we see that $r^2(z, y)/4s > r^2/2s + \delta/t$ for $s \in (0, t]$. Thus the last integral is bounded above by

$$
C e^{-\delta/t} \int_0^t s^{-(n/2)-1} \exp(-r^2/2s) \, ds,
$$

where this C equals to the old C times the volume of Ω. There exists a constant D such that $s^{-(n/2)-1} \leq D \exp(r^2/4s) \leq D \exp(r^2/4t)$ for $s \in (0, 1], s < t$, so the last integral is bounded above by

$$
C e^{-\delta/t} \int_0^t \exp(-r^2/4t) \, ds = C e^{-\delta/t} \exp(-r^2/4t) \cdot t.
$$

Since $t \leq D'(4\pi t)^{-n/2}$ for some constant D' for $t \in (0,1]$, we can bound this last line by

$$Ce^{-\delta/t}(4\pi t)^{-n/2}\exp(-r^2/4t).$$

Finally, since the function $u_0(x,y)$ is bounded away from zero on $B_r(x)$, this last expression is bounded above by

$$Ce^{-\delta/t}\bar{E},$$

which finishes the proof.

If we now substitute (4.22) into (4.21) and use this lemma, we obtain

$$\bar{e}(t,x,y) = e(t,x,y) + \int_0^t ds \int_{B_{\rho/4}(x)} e(t-s,x,z) \wedge *_z G(s,z,y)\bar{e}(s,z,y)$$
$$+ Fe^{-\delta/t}\bar{E}(t,x,y),$$

for some bounded function F. Note that $H = G$ on $B_{\rho/4}$ and that the last integral can be taken over all of M, due to the presence of ϕ in H. For $t \leq 1$, the last term in this equation is bounded above by a constant times $t^N \bar{E}$ for any $N > 0$. Since $\bar{E}(t,x,y)$ is $O(t^N)$ for any N if $x \neq y$ and is $O(t^{-n/2})$ for $x = y$, this last term does not contribute to the asymptotics of $e - \bar{e}$ as $t \to 0$.

Recall that any endomorphism A of $\Lambda^* T_x^* M$ has the decomposition $A = c^{IJ} A_{IJ}$ in the notation of (2.28). We define the ℓth degree of A to be

$$A_\ell = \sum_{|I|+|J|=\ell} c^{IJ} A_{IJ};$$

this notation extends the notation $e(t,x,x)_{2n}$ defined previously. Now the isomorphism $P : \Lambda^* T_x^* M \to \Lambda^* T_y^* M$ given by parallel translation yields an isomorphism

$$\text{Hom}(\Lambda^* T_x^* M, \Lambda^* T_y^* M) \cong \text{End}(\Lambda^* T_x^* M),$$
$$B \mapsto P^{-1} \circ B,$$

and so we can define B_ℓ to be $(P^{-1} \circ B)_\ell$. In particular,

$$P(x,y)_\ell = \begin{cases} \text{Id}, & \ell = 0, \\ 0, & \ell \neq 0. \end{cases}$$

Note that for the composition of endomorphisms, we have

$$(AB)_\ell = \sum_{a+b=\ell} A_a B_b.$$

Using Lemma 4.15, we find

$$\|(\bar{e} - e)_{2n}\| \leq \left\| \int_0^t ds \int_{B_{\rho/4}(x)} \sum_{a+b=2n} e(t-s,x,z)_a \right.$$

$$\wedge *_z (G(s, z, y)\bar{e}(s, z, y))_b \Big\|$$

$$+|F e^{-\delta/t} \bar{E}(t, x, y)|$$

$$\leq \left\| \int_0^t ds \int_{B_{\rho/4}(x)} \sum_{a+b=2n} \bar{e}(t - s, x, z)_a \right.$$

$$\wedge *_z (G(s, z, y)\bar{e}(s, z, y))_b \Big\|$$

$$+ \left\| \int_0^t ds \int_{B_{\rho/4}(x)} (t - s) \sum_{a+b=2n} e(t - s, x, z)_a \right.$$

$$\wedge *_z (G(s, z, y)\bar{e}(s, z, y))_b \Big\|$$

$$+|F e^{-\delta/t} \bar{E}(t, x, y)|.$$

To prove Theorem 4.11, we can ignore the last term in this inequality. Comparing the two integrals, we see that the second integral has an additional power of t coming from the $t - s$ term. Thus it suffices to show that

$$\left\| \int_0^t ds \int_{B_{\rho/4}(x)} \sum_{a+b=2n} \bar{e}(t - s, x, z)_a \wedge *_z (G(s, z, y)\bar{e}(s, z, y))_b \right\| \leq Ct. \quad (4.25)$$

The first step is to estimate $\bar{e}_a$. As previously noted,

$$B(t, y) = \int_0^t P^{-1} R P(x, \gamma(s)) = B(0, y) + t R(x) + O(t^2) = O(t). \quad (4.26)$$

By Exercise 7, as a nilpotent endomorphism of $\Lambda^* T^*_{\gamma(s)} M$, R adds in two a_i's and two a_i^*'s, as does the nilpotent endomorphism B. We have

$$(e^{-B})_a = \left(\sum_k \frac{(-1)^k (\int_0^t P^{-1} R P)^k}{k!} \right)_a. \quad (4.27)$$

Clearly the first term on the right hand side of (4.27) that can contribute a of each of the a_i, a_i^* occurs at k the first integer with $k \geq (a/4)$, and by (4.26), this contribution is $O(t^{a/4})$. As in (4.14), this may not be the only term contributing to $(e^{-B})_a$. However, these higher contributions are all $O(t^{1+(a/4)})$, and so we conclude that $(e^{-B})_a = O(t^{a/4})$. Since $\bar{E}$ is a function and hence has $\bar{E} = (\bar{E})_0$, we obtain

$$\|\bar{e}(t, x, y)_a\| = \|\bar{E}(t, x, y)(P(x, y) \cdot e^{-B})_a\| < C\bar{E}(t, x, y) t^{a/4} \quad (4.28)$$

for some constant $C > 0$.

The computation of $(G\bar{e})_b$ proceeds similarly. For example, one term in $(G\bar{e})_b$ equals

$$\sum_{e+f+g=b} (\nabla^* \nabla (e^{-B} P))_e (P^{-1} e^B)_f \bar{e}_g$$

$$= \sum_{e+f+g=b} O(s^{e/4})O(s^{f/4})O(s^{g/4})\bar{E}(s,z,y)$$

$$= O(s^{b/4})\bar{E}(s,z,y).$$

For as in Exercise 7, $\nabla^*\nabla e^{-B}$ is a complicated expression involving covariant derivatives of B. By that exercise, each covariant derivative term contributes two a_i's and two a_i^*'s. All other terms in $(G\bar{e})_b$ are also $O(s^{b/4})\bar{E}(s,z,y)$, with one exception: the term

$$(\frac{1}{2t}\langle\nabla r^2, \nabla e^{-B}P\rangle P^{-1}e^B\bar{e})_b = O(t^{(b/4)-1}). \tag{4.29}$$

Ignoring this term for the moment, we can estimate (4.25) by

$$\left\|\int_0^t ds \int_{B_{\rho/4}(x)} \sum_{a+b=2n} \bar{e}(t-s,x,z)_a \wedge *_z(G(s,z,y)\bar{e}(s,z,y))_b\right\| \tag{4.30}$$

$$\leq \left|\int_0^t ds \int_{B_{\rho/4}(x)} \sum_{a+b=2n} O((t-s)^{a/4})O(s^{b/4})\bar{E}(t-s,x,z) \wedge *_z\bar{E}(s,z,y)\right|.$$

To proceed, we prove a simple estimate on $\bar{E}$. Since the heat kernel on functions satisfies $e(t,x,y) \geq 0$ by Theorem 3.31, we have

$$\left|\int_{B_{\rho/4}(x)} \bar{E}(t-s,x,z)\bar{E}(s,z,y)\mathrm{dvol}_z\right|$$

$$= \left|\int_{B_{\rho/4}(x)} [e(t-s,x,z)(1+O(t-s))]\right.$$

$$\left. \times[e(s,z,y)(1+O(s))]\mathrm{dvol}_z\right|$$

$$= \left|\int_{B_{\rho/4}(x)} e(t-s,x,z)e(s,z,y)\mathrm{dvol}_z\right| + O(t)$$

$$\leq \int_M e(t-s,x,z)e(s,z,y)\mathrm{dvol}_z + O(t)$$

$$= e(t,x,y) + O(t)$$

$$= \bar{E}(t,x,y)(1+O(t)) + O(t)$$

$$\leq 2\bar{E}(t,x,y), \tag{4.31}$$

for small t. Here we have used Lemma 4.15, (4.19) and the trivial estimate $t-s \leq t$.

Except for the term in (4.29), (4.30) is thus bounded above by a constant times

$$\bar{E}(t,x,y)\int_0^t ds \sum_{a+b=2n} O((t-s)^{a/4})O(s^{b/4}) = \bar{E}(t,x,y)O(t^{(n/2)+1}) = O(t).$$

$$\tag{4.32}$$

Therefore (4.25) will be shown once we handle the exceptional term (4.29). For fixed x, this term is $O(r)$ in $r = r(x, y)$, since $\nabla r^2 = 2r dr$ (except at $r = 0$ where r is not differentiable). Since it also has a $1/s$ term in it, (4.29) contributes a term to the estimate (4.30) of the form

$$\int_0^t ds \int_{B_{\rho/4}(x)} \bar{E}(t - s, x, z)\bar{E}(s, z, y)O(r)O((t - s)^{a/4})O(s^{(b/4)-1})\ \mathrm{dvol}_z$$

$$= \int_0^t ds\ O(t^{(n/2)-1}) \int_{B_{\rho/4}(x)} O(r)\bar{E}(t - s, x, z)\bar{E}(s, z, y)\ \mathrm{dvol}_z. \quad (4.33)$$

(The $O(s^{(b/4)-1})$ term comes from

$$\frac{1}{s}(\langle\nabla r, \nabla e^{-B}P\rangle P^{-1}e^B \bar{e})_b = \frac{1}{s}\sum_{c+d=b}(\nabla_r e^{-B} \cdot e^B)_c \bar{e}_d.$$

$(\nabla_r e^{-B} \cdot e^B)_c$ involves $\nabla_r B$ which is treated as in Exercise 7.)

We now set $x = y$ and estimate (4.33). Since $\bar{E}$ is positive, it suffices to estimate

$$\int_{B_{\rho/4}(x)} e^{-r^2(x,z)/4(t-s)}r(x, z)e^{-r^2(z,x)/4s}\ \mathrm{dvol}_z.$$

In Riemannian normal coordinates centered at x, the metric is given by $g_{ij}(z) = \delta_{ij} + O(|z|^2)$, so for small ρ we can estimate the last integral by a constant times

$$\int_{|z|<\alpha} e^{-|z|^2/4(t-s)}|z|e^{-|z|^2/4s}\ dz, \quad (4.34)$$

where this integral is in $\mathbf{R}^n$ and α is some constant.

We claim that this integral is $O(t^{(n+1)/2})$ for n even. Converting to polar coordinates, we find that (4.34) is bounded above by a constant times

$$\int_0^\alpha r^n e^{-t|r|^2/4s(t-s)}\ dr = \frac{1}{2}\int_{-\alpha}^\alpha r^n e^{-t|r|^2/4s(t-s)}\ dr.$$

For $n = 0$, an explicit integration (from $-\infty$ to ∞) shows that the integral is $O(t^{1/2})$. For odd n, the right hand integral vanishes. For even n, writing $r^n = r^{n-1}r$ and doing an integration by parts gives

$$\int_{-\alpha}^\alpha r^n e^{-t|r|^2/4s(t-s)}\ dr = \frac{2(n - 1)s(t - s)}{t}\int_{-\alpha}^\alpha r^{n-2}e^{-t|r|^2/4s(t-s)}\ dr$$

$$= O(t^{(n+1)/2})$$

by induction.

Plugging this estimate for (4.34) into (4.33), we see that the exceptional term (4.29) contributes

$$\int_0^t ds\ O(t^{(n/2)-1})O(t^{(n+1)/2}) = O(t^{n+(1/2)}).$$

This completes the proof of Theorem 4.11.

Since the key to this hard technical argument is guessing the right approximate heat kernel (4.10), we should motivate its appearance. Of course, by 1944 it was known that the integrand in the Chern-Gauss-Bonnet theorem must be the Euler form. The fermion calculus was developed in the early 1980s (although an earlier version was known to physicists as the Berezin integral, and one can argue that Patodi's proof implicitly uses fermion calculus). After the crucial insight that only $e(t, x, x)_{2n}$ contributes to the integrand, it is easy to show, as we did, that replacing $e(t, x, x)_{2n}$ with $\bar{e}(t, x, x)_{2n}$ gives the Euler form. So some clever guesswork could lead to conjecturing Theorem 4.11.

Alternatively, one can argue as follows. The interpretation of heat flow as Brownian motion leads to the construction of a family of measures μ_t, the Wiener measures, on the space of continuous functions $\mathcal{C} = C([0, t], M)$ from $[0, t]$ to M with fixed value x_0 at $t = 0$. For $x \in \mathcal{C}$, it is standard to denote $x(t)$ by x_t. If $\mathbf{E}_t$ denotes expectation with respect to μ_t, then we have

$$e^{-t\Delta}f(x) = \mathbf{E}(f(x_t))$$

for $f \in C^\infty(M)$. Moreover, if $V: M \to \mathbf{R}$ is a smooth function, then the Schrödinger operator $\square = \Delta + V$ has a heat semigroup given by the famous Feynman-Kac formula

$$e^{-t\square}f(x_0) = \mathbf{E}(f(x_t)e^{\int_0^t V(x_s)\,ds}).$$

On forms, both the Bochner Laplacian $\nabla^*\nabla$ and Δ^k generalize the Laplacian on functions. From the Brownian motion point of view, the Bochner Laplacian is more natural, in that the Weitzenböck formula $\Delta^k = \nabla^*\nabla + R$ yields a Feynman-Kac formula

$$e^{-t\Delta^k}\omega_{x_0}(v_1, \ldots, v_k) = \mathbf{E}(\omega_{x_t}(v_1^t, \ldots, v_k^t)),$$

where v_t^j satisfies the covariant differential equation $\nabla_{\dot{x}_t} v_t^j = P_{x_t, x_0} R_{x_t} v_t^j$. (Since x is not differentiable in general, this must be interpreted as a stochastic integral equation.) Thus formally, $v_t^j = \exp(\int_0^t P_{x_s, x_0} R_{x_s} v_{x_s}^j\,ds)$. A reference for this material is [23].

The formulas for the heat operators on functions and forms differ superficially by the term $\exp(\int_0^t P_{x_s, x_0} R_{x_s} v_{x_s}^j\,ds)$. One might expect that for short time, heat tends to flow along geodesics; precise statements and proofs are in [49], [52]. Thus we can hope to replace $\int_0^t P_{x_s, x_0} R_{x_s} v_{x_s}^j\,ds$ just by an integral along the geodesic from x_0 to x_t, at least to O(t). Since we are interested in $e(t, x, x)$, the term $u_0(x, y)$ in $\bar{e}(t, x, y)$ is just one. Putting this all together, we conclude (with perhaps little confidence) that $\bar{e}(t, x, x)$ should approximate $e(t, x, x)$ fairly well for small time.

4.2   The Hirzebruch Signature Theorem and the Atiyah-Singer Index Theorem

4.2.1   A Survey of Characteristic Forms

In Chapter 1, we saw how to construct de Rham cohomology classes analytically via the Hodge theorem. In this subsection, we give a quick overview of Chern-Weil theory, which produces cohomology classes directly from the curvature of a Riemannian metric. At the end, we outline how these constructions carry over from the Levi-Civita connection on TM to connections on vector bundles.

Let M be a connected, closed, oriented n-manifold as usual. We claim that the map $\int : H_{dR}^n(M) \to \mathbf{R}$ given by $[\omega] \to \int_M \omega$ is an isomorphism (see Chapter 1, Exercise 33). The map is well defined by Stokes' theorem, and is surjective since $\int_M d\mathrm{vol}_g > 0$ for any metric g. Since $H_{dR}^n(M) \cong \mathbf{R}$ by Corollary 1.47, the map must be an isomorphism. As a result, two n-forms are cohomologous iff their integrals over M are identical.

In particular, the Euler forms $\omega_g, \omega_{g'}$ are in the same cohomology class by the Chern-Gauss-Bonnet theorem. Thus, in contrast to Hodge theory methods in which the metric is fixed, the Euler form is a curvature expression whose cohomology class is independent of the metric.

We now outline the Chern-Weil prescription for producing other expressions in the curvature whose cohomology class is independent of the metric. Given a local frame of TM, by which we mean a set $\{s_i(x)\}$ of n linearly independent vector fields defined over some neighborhood in M, we define the *connection one-forms* $\{\omega_j^i\}$ for the Levi-Civita connection ∇ by

$$\nabla s_j(x) = \omega_j^i(x) s_i(x).$$

For example, if $s_j = \partial_j$, then $\omega_j^i = \Gamma_{\ell j}^i dx^\ell$ by Definition 2, Chapter 2. Note that the ω_j^i are only defined locally, and depend on the choice of local frame. We also define the *curvature two-forms* by

$$\Omega_j^i = d\omega_j^i - \omega_k^i \wedge \omega_j^k.$$

We set $\Omega = (\Omega_j^i)$ to be the matrix of curvature two-forms and set $d\Omega = (d\Omega_j^i)$. We define multiplication of matrices of forms by the usual matrix multiplication rule where we wedge the entries:

$$(a_j^i) = (b_j^k)(c_k^i) \Leftrightarrow a_j^i = b_j^k \wedge c_k^i.$$

We similary define the Lie bracket of matrices of forms, where the Lie bracket for matrices is given by $[X, Y] = XY - YX$. In particular, we have by definition

$$\Omega = d\omega - \omega \wedge \omega = d\omega - \frac{1}{2}[\omega, \omega]. \tag{4.35}$$

Exercise 8: *(i) Show that if $\{s_j\}$ is an orthonormal frame with dual frame $\{\theta^j\}$ of one-forms, then*

$$\Omega_j^i = R_{jkl}^i \theta^k \wedge \theta^l,$$

with the curvature tensor components computed in this frame.

(ii) Show that if $\{s_j\}$ is an orthonormal frame, then $\Omega^i_j = -\Omega^j_i$. Hint: replace X, Y, Z by s_k, s_i, s_j in the first equation in (2.24).

(iii) Show that if we change from one frame to another, then Ω changes to $T^{-1}\Omega T$, for some invertible $n \times n$ matrix T. Conclude that Ω is a globally defined matrix of two-forms, i.e. Ω defines a linear transformation (on what space?) that is independent of coordinates.

(iv) Prove the Bianchi identity

$$d\Omega = [\omega, \Omega]$$

by differentiating (4.35).

Let $M_n(\mathbf{F})$ denote the set of $n \times n$ matrices with entries in $\mathbf{F}$, where $\mathbf{F}$ is $\mathbf{R}$ or $\mathbf{C}$. Let $P : M_n(\mathbf{F}) \to \mathbf{F}$ be an invariant polynomial in the entries of the matrices, i.e. $P(T^{-1}XT) = P(X)$ for all $X \in M_n(\mathbf{F})$ and invertible $T \in M_n(\mathbf{F})$. For example, we could set $P(X) = \text{Tr } X$, $P(X) = \det X$, or (for $\mathbf{F} = \mathbf{C}$) $P(X)$ could be the ith elementary symmetric polynomial in the eigenvalues of X. It is a standard result that these examples generate the algebra of invariant polynomials.

An invariant polynomial P can be evaluated on the matrix Ω, if it is understood that the multiplication operations in P are replaced by wedge product, to produce a form $P(\Omega) \in \bigoplus_k \Lambda^{2k} T^* M$. If we assume that P is homogeneous of degree q, then $P(\Omega) \in \Lambda^{2q} T^* M$. By Exercise 8, $P(\Omega)$ is independent of coordinates precisely because P is invariant. For example,

$$\text{Tr}(\Omega) = \Omega^1_1 + \Omega^2_2 + \ldots + \Omega^n_n.$$

However, this expression is zero since Ω is conjugate to a skew-symmetric matrix by the last exercise. The same vanishing occurs for the odd elementary symmetric polynomials, but the even polynomials need not vanish.

Lemma 4.36 *Assume that P is homogeneous of degree q.*

(i) $P(\Omega)$ is always closed.

(ii) If Ω_1 is the curvature matrix associated to a metric g_1 on M, then there exists a $(2q - 1)$-form $\theta = \theta(\Omega, \Omega_1)$ such that $P(\Omega) = P(\Omega_1) + d\theta$.

PROOF. (i) The coefficient $C(X_1, \ldots, X_q)$ of $t_1 \cdot \ldots \cdot t_q$ in $P(t_1 X_1 + \ldots + t_q X_q)$ is a symmetric multilinear function of $X_1, \ldots, X_q$. We call $(1/q!)C = p(X_1, \ldots, X_q)$ the complete polarization of P. It satisfies $p(X, \ldots, X) = P(X)$. This can be easily derived from P's Taylor expansion using

$$C = \frac{\partial^q}{\partial t_1 \cdot \ldots \cdot \partial t_q}\bigg|_{t_1 = \ldots = t_q = 0} P(t_1 X_1 + \ldots + t_q X_q).$$

The invariance of P implies

$$p(T^{-1}X_1 T, \ldots, T^{-1}X_q T) = p(X_1, \ldots, X_q). \tag{4.37}$$

Given $Y \in M_n$, let $T(t)$ be a one parameter family of invertible matrices with $T(0) = \text{Id}, \dot{T}(0) = Y$. Note that

$$\frac{d}{dt}\Big|_{t=0} T^{-1}XT = [X,Y].$$

Differentiating (4.37) gives

$$\sum_{i=1}^{q} p(X_1, \ldots, [X_i, Y], \ldots, X_q) = 0. \qquad (4.38)$$

But then

$$\begin{aligned} dP(\Omega) &= dp(\Omega, \ldots, \Omega) = qp(d\Omega, \ldots, \Omega) \\ &= qp([\Omega, \omega], \Omega, \ldots, \Omega) = 0, \end{aligned}$$

by the Bianchi identity and (4.38).

(ii) For $t \in [0,1]$, define a family of operators $\nabla_t : \Gamma(TM) \to T^*M \otimes \Gamma(TM)$ by $\nabla_t = (1-t)\nabla_0 + t\nabla_1$, where ∇_0, ∇_1 are the Levi-Civita connections for $g = g_0, g_1$, respectively. Equivalently, we have $\nabla_t = \nabla_0 + t\alpha$, for $\alpha = \nabla_1 - \nabla_0$.

Each ∇_t has an associated matrix (ω_t) of one-forms defined by $\nabla_t s_j = (\omega_t)^i_j s_i$; it is easy to check that $\omega_t = \omega_0 + t\alpha$, where we use α to denote also the matrix of one-forms for the operator $\nabla_0 - \nabla_1$. We define a "curvature" matrix by $(\Omega_t) = (d\omega_t - (1/2)[\omega_t, \omega_t])$ with $\Omega_0 = \Omega$.

A direct computation gives

$$\Omega_t = \Omega_0 + t(d\alpha - [\omega_0, \alpha]) - \frac{1}{2}t^2[\alpha, \alpha].$$

Thus

$$\begin{aligned} \frac{1}{q}\frac{d}{dt}P(\Omega_t) &= p(d\alpha - [\omega_0, \alpha] - t[\alpha, \alpha], \Omega_t, \ldots, \Omega_t) \\ &= p(d\alpha - [\omega_t, \alpha], \Omega_t, \ldots, \Omega_t). \end{aligned}$$

Now

$$dp(\alpha, \Omega_t, \ldots, \Omega_t) = p(d\alpha, \Omega_t, \ldots, \Omega_t) - (q-1)p(\alpha, [\omega_t, \Omega_t], \Omega_t, \ldots, \Omega_t).$$

Moreover, as in (4.38),

$$p([\omega_t, \alpha], \Omega_t, \ldots, \Omega_t) - (q-1)p(\alpha, [\omega_t, \Omega_t], \Omega_t \ldots, \Omega_t) = 0.$$

(The minus sign appears in the last equation because the first entry in p is a one-form rather than a two-form.) Combining the last three equations gives

$$\frac{1}{q}\frac{d}{dt}P(\Omega_t) = dp(\alpha, \Omega_t, \ldots, \Omega_t),$$

and so

$$P(\Omega_1) - P(\Omega_0) = q \int_0^1 dp(\alpha, \Omega_t, \ldots, \Omega_t) \, dt = d\left(q \int_0^1 p(\alpha, \Omega_t, \ldots, \Omega_t) \, dt \right).$$

Setting $\theta(\Omega, \Omega_1) = q \int_0^1 p(\alpha, \Omega_t, \ldots, \Omega_t) dt$ finishes the proof.

Thus to each invariant polynomial P, homogeneous of degree q, and each Riemannian metric g on M, we may associate a *characteristic form* $P(\Omega)$ such that the *characteristic class* $[P(\Omega)] \in H_{dR}^{2q}(M)$ is independent of g. For the even elementary symmetric polynomial P_{2i}, the associated cohomology class $(-1)^i [P_{2i}(\Omega)] \in H_{dR}^{4i}(M)$ is called the ith *Pontrjagin class* of M; any characteristic form $P(\Omega)$ is a universal polynomial in the $P_{2i}(\Omega)$.

Exercise 9: *Let $f : N \to M$ be an immersion. If g is a metric on M, show that f^*g is a metric on N. Show that for any invariant polynomial P, we have $[P(\Omega_{f^*g})] = f^*[P(\Omega_g)] \in H_{dR}^{2q}(N)$. Hint: first show that $\Omega_{f^*g} = f^*\Omega_g$.*

If we restrict attention to a subset of $M_n(\mathbf{R})$, we may find new examples of invariant polynomials. In particular, for n even consider the equation $P(T^{-1}XT) = P(X)$ for X a skew-symmetric $n \times n$ matrix and $T \in SO(n)$. There is exactly one new invariant P (up to multiplication by a constant), the *Pfaffian* $\mathrm{Pf}(X)$ of the matrix. The Pfaffian is a polynomial with integral coefficients characterized (up to sign) by the equation $(\mathrm{Pf}(X))^2 = \det X$. For example,

$$\mathrm{Pf}\begin{pmatrix} 0 & a \\ -a & 0 \end{pmatrix} = a, \tag{4.39}$$

and if (a_{ij}) is a skew-symmetric 4×4 matrix, then $\mathrm{Pf}(a_{ij}) = a_{12}a_{34} - a_{13}a_{24} + a_{14}a_{23}$. In particular, if $\dim M = 2$, then with respect to a local orthonormal frame $\{\theta^i\}$ of T^*M we have

$$(\Omega_j^i) = \begin{pmatrix} 0 & \Omega_2^1 \\ -\Omega_2^1 & 0 \end{pmatrix},$$

which implies by Exercise 8

$$\mathrm{Pf}(\Omega) = \Omega_2^1 = R_{2kl}^1 \theta^k \wedge \theta^l = 2R_{212}^1 \theta^1 \wedge \theta^2 = -2K \, dA.$$

(By the remark at the beginning of this section, the fact that $[\mathrm{Pf}(\omega)]$ is independent of the metric is thus equivalent to $\int_M K \, dA$ being independent of the metric, which finally furnishes an explanation for the remarkable nature of the Gauss-Bonnet theorem mentioned in Theorem 2.6.) In fact, the Euler form ω of §4.1 is precisely given by

$$\omega = \frac{(-1)^{n/2}}{(4\pi)^{n/2}} \mathrm{Pf}(\Omega),$$

as the next exercise shows.

Exercise 10: *(i) Let V be an oriented real vector space of dimension $n = 2m$ with positively oriented basis $\psi_1, \ldots, \psi_n$. Set $\psi = (\psi_1, \ldots, \psi_n)$. Let $\omega = (\omega_{ij})$ be a skew-symmetric matrix. Consider the two-form in $\Lambda^* V$ given by $\frac{1}{2}\omega_{jk}\psi_j \wedge \psi_k$, which we denote by $\frac{1}{2}\psi^t \omega \psi$. Define the Pfaffian of ω by*

$$\frac{1}{m!}\left(\frac{1}{2}\psi^t \omega \psi\right)^m = \mathrm{Pf}(\omega)\psi_1 \wedge \ldots \wedge \psi_n. \tag{4.40}$$

Show that $\mathrm{Pf}(\omega)$ is a homogeneous polynomial of degree m in the entries of ω, and verify (4.39).

(ii) Now assume that V has an inner product and that $\psi_1, \ldots, \psi_n$ and $\mu_1, \ldots, \mu_n$ are two positively oriented bases. Let $\omega: V \to V$ be a skew-adjoint transformation with matrices $(\omega_{ij}^\psi), (\omega_{ij}^\mu)$, with respect to the ψ and μ bases. Show that

$$\mathrm{Pf}(A\omega A^t) = \mathrm{Pf}(\omega)\det(A),$$

where A is the matrix expressing ψ_i in terms of μ_j. Conclude that $\mathrm{Pf}(\omega)$ is an $SO(n)$ invariant polynomial on skew-symmetric matrices. Show also that

$$\mathrm{Pf}(\omega^\psi)\psi_1 \wedge \ldots \wedge \psi_n = \mathrm{Pf}(\omega^\mu)\mu_1 \wedge \ldots \wedge \mu_n.$$

We denote either side of this equation by $\mathrm{Pf}(\omega)\mathrm{dvol}$, a well defined element of $\Lambda^n V$.

(iii) Complexify V to $V_{\mathbf{C}} = V \otimes_{\mathbf{R}} \mathbf{C}$. Consider the skew-Hermitian transformation $i\omega$ on $V_{\mathbf{C}}$. Check that its eigenvalues are all real, and that if v is an eigenvector, then so is its complex conjugate $\bar{v}$. Conclude that $V_{\mathbf{C}}$ has an orthonormal basis of eigenvectors $v_1, \bar{v}_1, \ldots, v_m, \bar{v}_m$ with eigenvalues denoted $\pm \lambda_1, \ldots, \pm \lambda_m$. Define real vectors

$$w_1 = \frac{v_1 + \bar{v}_1}{2}, w_2 = \frac{iv_1 - i\bar{v}_1}{2}, \text{ etc.}$$

Show that $\omega w_1 = -\lambda_1 w_2, \omega w_2 = \lambda_1 w_2$, etc. Conclude that there exists $B \in O(n)$ such that

$$B\omega B^t = \begin{pmatrix} 0 & \lambda_1 & & & & \\ -\lambda_1 & 0 & & & & \\ & & \ddots & & & \\ & & & & 0 & \lambda_m \\ & & & & -\lambda_m & 0 \end{pmatrix}.$$

Thus $\mathrm{Pf}(B\omega B^t) = \pm\mathrm{Pf}(\omega)$. Use (i) to show that $\mathrm{Pf}(\omega) = \lambda_1 \cdot \ldots \cdot \lambda_m$. Conclude that $\mathrm{Pf}(\omega)^2 = \det(\omega)$. Show that in fact $\mathrm{Pf}(\omega)$ is the positive square root of $\det(\omega)$ by considering the case $\lambda_1 = \ldots = \lambda_m = 1$.

(iv) Given an oriented Riemannian n-manifold M, choose a local positively oriented orthonormal frame of $T^* M$. By Exercise 8(ii), the corresponding curvature matrix Ω is skew-symmetric. Show that the n-form $\mathrm{Pf}(\Omega)$ is independent of choice of frame.

(v) Relabel $\psi_1, \ldots, \psi_m$ as $a_1^*, \ldots, a_m^*$ and $\psi_{m+1}, \ldots, \psi_n$ as $a_1, \ldots, a_m$ in the notation of §2.2.2. Show that

$$\mathrm{Tr}((-1)^F e^{(t/2)\psi^t \omega \psi}) = m! \mathrm{Pf}(\omega) t^m + \mathrm{O}(t^{m+1}).$$

Conclude that $\mathrm{Pf}(\Omega)$ equals the Euler form up to a factor of $(-1)^m (4\pi)^{-m}$.

Remark: We now assume familiarity with singular homology and cohomology. Define the *Euler class* $e = e(M)$ to be $[\omega] \in H_{dR}^n(M) \cong H_{sing}^n(M; \mathbf{R})$; a purely topological definition of the Euler class is in [47]. From the proof of the de Rham theorem mentioned in §1.4, it follows that for any closed n-form α,

$$\int_M \alpha = \langle [\alpha], [M] \rangle,$$

where $[\alpha]$ also denotes the singular cohomology class associated to the de Rham cohomology class of α, and $[M] \in H_n(M; \mathbf{R})$ is the fundamental homology class. We have

$$\chi(M) \stackrel{(1)}{=} \sum_q (-1)^q \dim \mathrm{Ker}\ \Delta^q \stackrel{(2)}{=} \int_M \omega \stackrel{(3)}{=} \langle e, [M] \rangle.$$

In this equation, (1) follows from choosing a Riemannian metric on M and using Hodge theory, (2) uses the heat equation method, and (3) uses a combination of Chern-Weil theory and the de Rham theorem. The final form of our generalization of the Gauss-Bonnet theorem is thus a theorem in differential topology,

$$\chi(M) = \langle e, [M] \rangle,$$

proven in our case by geometric means.

Much of this section generalizes to connections on vector bundles. Here a *connection* ∇ on a real or complex vector bundle E is a map $\nabla : \Gamma(E) \to \Gamma(E \otimes T^*M)$ satisfying $\nabla(fs) = df \cdot s + f \nabla s$, for all $s \in \Gamma(E), f \in C^\infty(M)$ (cf. §2.2, Definition 2). If $\{s_j\}$ is a locally defined set of sections such that $\{s_j(x)\}$ forms a basis of the fiber E_x at each x in the domain of the s_j, then $\nabla s_j = \omega_j^i s_i$ for some matrix (ω_j^i) of one-forms. We call the matrix of two-forms (Ω_j^i) defined by $\Omega_j^i = d\omega_j^i - \omega_k^i \wedge \omega_j^k$ the *curvature* of the connection ∇.

Exercise 11: *(i) Show that the curvature is independent of the choice of the $\{s_j\}$, and is thus a globally defined matrix of two-forms.*

(ii) Let ∇ be the Levi-Civita connection for a metric g on M. Replace E above by TM, and show that the two definitions of the curvature matrix of the Levi-Civita connection coincide. Hint: let $s_j = \partial_{x^j}$.

Now assume that E is a complex vector bundle. For any invariant polynomial P, homogeneous of degree q, we have as before that $[P(\Omega)] \in H_{dR}^{2q}(M)$ is independent of the choice of connection on E, and thus provides a topological invariant $P(E)$ of the vector bundle. For the elementary symmetric polynomial P_i, $P_i(E)$ is called the ith *Chern class* of E.

Exercise 12: *Let E, F be vector bundles over M, and let $f : E \to F$ be a smooth bundle map. If ∇ is a connection on F, define $f^* \nabla$ and show that it is a connection on E. If P is an invariant polynomial, show that $f^* P(F) = P(f^* F)$.*

4.2.2   The Hirzebruch Signature Theorem

In this subsection we will outline a heat equation proof that the signature, a subtler topological invariant than the Euler characteristic, is also given by the integral of a characteristic form. As in the last subsection, this result can be interpreted as a purely topological statement.

Assume $\dim M = 4k$. Define the *intersection pairing*

$$\cdot : H^{2k}_{dR}(M) \otimes H^{2k}_{dR}(M) \to \mathbf{R}$$

by

$$[\omega] \cdot [\eta] = \int_M \omega \wedge \eta.$$

This is a symmetric bilinear form and is well defined by Stokes' theorem:

$$
\begin{aligned}
\int_M (\omega + d\theta) \wedge \eta &= \int_M \omega \wedge \eta + \int_M d\theta \wedge \eta \\
&= \int_M \omega \wedge \eta + \int_M d(\theta \wedge \eta) - \int_M \theta \wedge d\eta \\
&= \int_M \omega \wedge \eta,
\end{aligned}
$$

as $d\eta = 0$. Moreover, the intersection pairing is nondegenerate, i.e. if $[\omega] \cdot [\eta] = 0$ for all $[\eta]$, then $[\omega] = 0$.

Exercise 13: *Prove this nondegeneracy. Hint: Pick a harmonic form $\alpha \in [\omega]$. Show that $*\alpha$ is also harmonic, and that $[\alpha] \cdot [*\alpha] > 0$, if $[\omega] \neq 0$.*

Thus there exists a basis $\{[\omega_i]\}$ of $H^{2k}_{dR}(M)$ such that the matrix $([\omega_i] \cdot [\omega_j])$ of the intersection pairing diagonalizes to

$$
\begin{pmatrix}
1 & & & & & \\
& \ddots & & & & \\
& & 1 & & & \\
& & & -1 & & \\
& & & & \ddots & \\
& & & & & -1
\end{pmatrix},
\tag{4.41}
$$

with, say, p plus ones and q minus ones.

Remark: For the reader familiar with singular cohomology theory, we can define the intersection pairing on the free part of $H^{2k}_{sing}(M;\mathbf{Z})$ using the cup product, the singular cohomology analogue of the wedge product in de Rham cohomology. The matrix of the intersection pairing with respect to any $\mathbf{Z}$-basis of the free part of $H^{2k}_{sing}(M;\mathbf{Z})$ will then have integer coefficients. However, there exist topological four-manifolds M whose intersection pairing matrix cannot be diagonalized over $\mathbf{Z}$, i.e. there exists no change of basis transformation with *integral* coefficients taking the intersection pairing matrix to the form in (4.41). Donaldson applied deep geometric techniques to show that if this M possessed a smooth structure, then the matrix would have to diagonalize via such an integral transformation ([22],[25]). This produced the first four-dimensional example of a topological manifold admitting no smooth structure. No such examples are possible in dimensions one and two. Whether such examples can exist in dimension three is among the most important unsolved problems in topology.

We set the *signature* $\sigma(M)$ of M to be $p - q$, the number of plus ones minus the number of minus ones in the diagonalized matrix. The signature is an invariant of the smooth structure of M, and by de Rham's theorem it is in fact a topological invariant. As for the Chern-Gauss-Bonnet theorem, we will find an analytic method for computing the signature in terms of a Riemannian metric, and then use the heat equation method to evaluate the signature.

We set the *index* of an operator $D : \mathcal{H} \to \mathcal{H}$ on a Hilbert space $\mathcal{H}$ to be

$$\mathrm{index}(D) = \dim \mathrm{Ker}\ D - \dim \mathrm{Ker}\ D^*,$$

provided both numbers on the right hand side are finite.

Exercise 14: *Show that $\chi(M)$ equals the index of*

$$d + \delta : \bigoplus_k L^2 \Lambda^{2k} T^* M \to \bigoplus_k L^2 \Lambda^{2k+1} T^* M.$$

Let $\Lambda^* = \bigoplus_q L^2 \Lambda^q T^* M$ be the space of *complex* valued forms on M^{4k}, and define an endomorphism $\tau : \Lambda^* \to \Lambda^*$ by $\tau|_{L^2 \Lambda^q T^* M} = i^{q(q-1)+k} *$, so that $\tau^2 = 1$. This yields a decomposition $\Lambda^* = \Lambda^+ \oplus \Lambda^-$ into the (sections of) bundles $\Lambda^{\pm} = \{\omega \in \Lambda^* ; \tau\omega = \pm\omega\}$ of the ± 1 eigenspaces of τ. (We are being sloppy in not distinguishing between a bundle and its sections.) Set $d + \delta = \bigoplus_q (d^q + \delta^q)$ acting on forms of mixed degree, and let $D = (d+\delta)|_{\Lambda^+}$. D is called the *signature operator* on M.

Exercise 15: *(i) Show that $D : \Lambda^+ \to \Lambda^-$ and that $D^* = d + \delta : \Lambda^- \to \Lambda^+$.*
(ii) Show that if ω is a harmonic form in $L^2 \Lambda^q T^* M$, then $\omega \pm \tau\omega \in \mathrm{Ker}\ D|_{\Lambda^{\pm}}$. Conclude that for $q \neq 2k$ the map $\omega + \tau\omega \to \omega - \tau\omega$ gives an isomorphism

$$\mathrm{Ker}\ D|_{\Lambda^+ \cap (\mathrm{Ker}\ \Delta^q \oplus \mathrm{Ker}\ \Delta^{4k-q})} \xrightarrow{\cong} \mathrm{Ker}\ D^*|_{\Lambda^- \cap (\mathrm{Ker}\ \Delta^q \oplus \mathrm{Ker}\ \Delta^{4k-q})}.$$

By the last exercise, we see that

$$\begin{aligned}\text{index}(D) &= \text{index}(D|_{\Lambda^+ \cap \Lambda^{2k}}) = \dim \text{Ker}\,(D|_{\Lambda^+ \cap \Lambda^{2k}}) - \dim \text{Ker}\,(D|_{\Lambda^- \cap \Lambda^{2k}})\\ &= \dim(\Lambda^+ \cap \text{Ker}\,(d+\delta)|_{\Lambda^{2k}}) - \dim(\Lambda^- \cap \text{Ker}\,(d+\delta)|_{\Lambda^{2k}}).\end{aligned}$$

For $\omega \in \Lambda^+ \cap \text{Ker}\,(d+\delta)|_{\Lambda^{2k}}$, we have $*\omega = \omega$ (as $\tau = *$ on Λ^{2k}), so

$$[\omega] \cdot [\omega] = \int_M \omega \wedge \omega = \int_M \omega \wedge *\omega > 0.$$

Similarly, for $\omega \in \Lambda^- \cap \text{Ker}\,(d+\delta)|_{\Lambda^{2k}}$, we have $[\omega] \cdot [\omega] < 0$. Since

$$H_{dR}^{2k} \cong \text{Ker}\,(d+\delta)|_{\Lambda^{2k}} = (\Lambda^+ \cap \text{Ker}\,(d+\delta)|_{\Lambda^{2k}}) \oplus (\Lambda^- \cap \text{Ker}\,(d+\delta)|_{\Lambda^{2k}}),$$

it follows that

$$\text{index}(D) = \sigma(M).$$

Thus, in analogy to Exercise 14, we have an analytic expression for the topological invariant $\sigma(M)$.

Exercise 16: *Show that $\sigma(M \times N) = \sigma(M)\sigma(N)$. Hint: Put a product metric on $M \times N$ and express $d_{M \times N}, \delta_{M \times N}$ in terms of $d_M, d_N, \delta_M, \delta_N$. Let $\{\alpha_i\}, \{\beta_j\}$ be bases of Λ_M^+, Λ_M^-, respectively, and let $\{\eta_i\}, \{\mu_j\}$ be bases for Λ_N^+, Λ_N^-, respectively. Show that $\Lambda_{M \times N}^+$ has basis $\{\alpha_i \otimes \eta_j, \beta_i \otimes \mu_j\}$, and that $\Lambda_{M \times N}^-$ has basis $\{\alpha_i \otimes \mu_j, \beta_i \otimes \eta_j\}$; see Exercise 47, Chapter 1.*

The next step is to convert this analytic expression into a geometric one. Define Laplacians

$$\begin{aligned}\Delta^+ &= D^*D = (\delta+d)(d+\delta) = \delta d + d\delta : \Lambda^+ \to \Lambda^+,\\ \Delta^- &= DD^* = \delta d + d\delta : \Lambda^- \to \Lambda^-.\end{aligned}$$

By a straightforward modification of the construction of the heat kernel for forms, it can be shown that $\Delta^\pm$ have heat kernels $e^\pm(t,x,y)$ and associated heat operators $e^{-t\Delta^\pm}$. As a result, the Hodge theory techniques of Chapter 1 yield an eigenform decomposition for $\Lambda^\pm$ with respect to $\Delta^\pm$. Similarly, if $E_\lambda^\pm$ is the λ-eigenspace for $\Delta^\pm$, then $D : E_\lambda^+ \to E_\lambda^-$ is an isomorphism if $\lambda \neq 0$. Thus

$$\begin{aligned}\sigma(M) &= \text{index}(D) = \text{Tr}\,e^{-t\Delta^+} - \text{Tr}\,e^{-t\Delta^-}\\ &= \frac{1}{(4\pi t)^{2k}} \int_M \left(\text{tr}\,a_{2k}^+(x,x) - \text{tr}\,a_{2k}^-(x,x) \right) \text{dvol},\end{aligned}$$

where

$$e^\pm(t,x,x) \sim \frac{1}{(4\pi t)^{2k}} \sum_{k=0}^\infty a_k^\pm(x,x)t^k.$$

Even without identifying the form $\text{tr}\,a_{2k}^+ - \text{tr}\,a_{2k}^-$, we can still obtain information about the signature. Given oriented $4k$-manifolds M_1, M_2, we form the connected sum $M_1 \# M_2$ by deleting $4k$-disks D_i from M_i, and then identifying one boundary component of the cylinder $C = S^{4k-1} \times [0,1]$ with $\partial D_1 \approx S^{4k-1}$ and the other boundary component with ∂D_2. With a little care, the resulting manifold is oriented.

Corollary 4.42 $\sigma(M_1 \# M_2) = \sigma(M_1) + \sigma(M_2)$.

PROOF. Put metrics g_i on M_i By a partition of unity argument, we may put a metric $\tilde{g}$ on $M_1 \# M_2$ which agrees with g_i on $M_i' = M_i - D_i$. If we denote $\operatorname{tr} a_{2k}^+(x, x) - \operatorname{tr} a_{2k}^-(x, x)$ by $L_k(x)$, then $L_k(x)$ for the metric $\tilde{g}$ agrees with $L_k(x)$ for the metrics g_i, for $x \in M_i'$, since $L_k(x)$ is computed by a local expression in the curvature of the metric and its covariant derivatives. Thus

$$
\begin{aligned}
\sigma(M_1 \# M_2) &= \int_{M_1 \# M_2} L_k = \int_{M_1'} L_k + \int_C L_k + \int_{M_2'} L_k \\
&= \int_{M_1} L_k - \int_{D_1} L_k + \int_C L_k + \int_{M_2} L_k - \int_{D_2} L_k \\
&= \sigma(M_1) + \int_{S^{4k}} L_k + \sigma(M_2),
\end{aligned}
$$

since (taking care for the orientations) $S^{4k} = D_1 \cup C \cup D_2$. Now $H^{2k}(S^{4k}) = 0$, so $\int_{S^{4k}} L_k = \sigma(S^{4k}) = 0$ for any metric on the sphere. (Note that the smoothing process used to form $\tilde{g}$ ensures that g_i and $\tilde{g}$ patch together to form a smooth metric on the sphere.) Thus $\sigma(M_1 \# M_2) = \sigma(M_1) + \sigma(M_2)$.

Exercise 17: *(i) If M' is an ℓ-fold cover of M, show that $\sigma(M') = \ell \cdot \sigma(M)$.*
(ii) Compute $\chi(M \# N)$ in terms of $\chi(M), \chi(N)$.

We now identify the integrand $\operatorname{tr} a_{2k}^+ - \operatorname{tr} a_{2k}^-$ in the formula for the signature. Once again, this expression *a priori* contains many covariant derivatives of the curvature, but in fact turns out to be an invariant polynomial in the curvature, by the work of Patodi [55], Gilkey [29], Atiyah, Bott and Patodi [1], and Getzler [28]. The supersymmetric derivation of this polynomial is more difficult than in the Chern-Gauss-Bonnet case, since the square of the signature operator does not have a good Weitzenböck decomposition (see [5], [59]).

To define this polynomial, called the *Hirzebruch L polynomial*, it is enough to define it on a diagonal matrix. We first define a formal Taylor series of the entries by

$$
\begin{aligned}
L_k &\begin{pmatrix} \lambda_1 & & \\ & \ddots & \\ & & \lambda_{4k} \end{pmatrix} \\
&= \prod_{i=1}^{4k} \frac{\lambda_i}{\tanh \lambda_i} \\
&= \prod_{i=1}^{4k} \left(1 + \frac{1}{3}\lambda_i^2 - \frac{1}{45}\lambda_i^4 + \ldots + (-1)^{q-1} 2^{2q} B_q \frac{\lambda_i^{2q}}{2q!} + \ldots \right)
\end{aligned}
$$

where B_q is the qth Bernoulli number. We now define the L_k polynomial to be the component in degree $2k$ of the Taylor series for L_k. When evaluated on a curvature matrix, this component is denoted $L_k(\Omega)_{4k}$.

Exercise 18: *Let $(M, g), (N, h)$ be manifolds of dimension $4k, 4\ell$, respectively, and let $g \oplus h$ denote the product metric on $M \times N$. Show that $L_{k+\ell}(\Omega_{g \oplus h}) = L_k(\Omega_g) L_\ell(\Omega_h)$.*

Theorem 4.43 (Hirzebruch Signature Theorem)

$$\sigma(M) = \int_M L_k(\Omega)_{4k}.$$

Exercise 19: *Compute $\sigma(T^4)$ in three ways: (i) directly from its definition, (ii) by finding the ± 1 eigenspaces of $*$ on $H^2_{dR}(T^4)$, and (iii) by the Hirzebruch signature theorem. Hint: In the last two parts, use the flat metric.*

As with the Chern-Gauss-Bonnet theorem, the signature theorem admits a purely topological formulation via Chern-Weil theory and de Rham's theorem. Since the $L_k(\Omega)_{4k}$ is a symmetric polynomial of even degree in the entries of the curvature matrix, it is a polynomial in the even degree elementary symmetric polynomials, and hence a polynomial in the Pontrjagin forms. For example, if we denote the ith Pontrjagin form by $p_i = p_i(\Omega)$, then

1. $L_1 = \frac{1}{3}p_1$,

2. $L_2 = \frac{1}{45}(7p_2 - p_1^2)$ (where $p_1^2 = p_1 \wedge p_1$),

3. $L_4 = \frac{1}{14175}(381p_4 - 71p_3p_1 - 14p_2^2 + 22p_2p_1^2 - 3p_1^4)$.

It is a good exercise to work out at least L_1, L_2. In particular, we get nontrivial integrality results. From

$$\sigma(M^4) = \frac{1}{3} \int_M p_1(\Omega),$$

we must have $\int_{M^4} p_1 \in 3\mathbf{Z}$. (For us, $\int_{M^4} p_1$ is *a priori* only a real number, but a more careful treatment of Chern-Weil theory shows that the integral of any polynomial of degree $4k$ in the Pontrjagin forms over a $4k$-manifold must be an integer. Here we get the added information that $\int_{M^4} p_1$ is divisible by 3.) Similarly,

$$\int_{M^{16}} (381p_4 - 71p_3p_1 - 14p_2^2 + 22p_2p_1^2 - 3p_1^4) \in 14175\mathbf{Z}.$$

Exercise 20: *(i) Let M, N be oriented cobordant manifolds, i.e. there exists an oriented $(4k+1)$-manifold W such that ∂W is the disjoint union of M and $-N$,* *where $-N$ means N with its orientation reversed. Show that $\sigma(M) = \sigma(N)$. Hint: Put a metric g on W such that in a collar $M \times [0,1]$ of M, g is a product metric, i.e. $g = g_M \oplus dt^2$, where g_M is a metric on M and $dt^2 = dt \otimes dt$ is the standard metric on $[0,1]$. Make sure that g is also a product metric $g_N \oplus dt^2$ on a collar of N. If $i : M \to W$ is the inclusion, show that $i^* \Omega_g = \Omega_{g_M}$, and similarly for N. Conclude from Exercise 9 that $i^* L_k(\Omega_g)_{4k} = L_k(\Omega_{g_M})_{4k}$. Thus*

$$\sigma(M) - \sigma(N) = \int_M L_k(\Omega_{g_M})_{4k} - \int_N L_k(\Omega_{g_N})_{4k} = \int_W dL_k(\Omega_g)_{4k} = 0.$$

(ii) Note that every oriented closed two-manifold is the boundary of a three-manifold, and hence is cobordant to the empty set. Why can't we modify the previous argument to conclude that the Euler characteristic of a closed oriented surface is zero?

Remark: The signature theorem, and the companion Hirzebruch-Riemann-Roch theorem in the next subsection, are striking because of the complicated coefficients of the characteristic forms. As motivation for the appearance of the L polynomial, it follows from cobordism theory that since the signature is unchanged under oriented cobordism, and is multiplicative for product manifolds, the signature must be given by a polynomial in the Pontrjagin classes. (Of course, these properties of the signature have purely topological proofs.) The signature of the complex projective spaces $\mathbf{CP}^{2k}$ is easily seen to be one. Thus the L polynomial must (i) be multiplicative (cf. Exercise 16), (ii) be a cobordism invariant (which holds for any polynomial in the Pontrjagin classes as in Exercise 20), and (iii) have $\int_{\mathbf{CP}^{2k}} L_k(\Omega_g) = 1$. Properties (i), (ii) show that L_k on a diagonal matrix must be the product of an even function of the entries, and property (iii) determines that even function. Hirzebruch's original topological proof of the signature theorem along these lines is in [47].

4.2.3 The Atiyah-Singer Index Theorem

Before stating the Atiyah-Singer index theorem, we give two more examples of operators for which the heat equation method yields results along the lines of the signature theorem. First, let E be a Hermitian vector bundle over a Riemannian manifold (M^{2k}, g) with compatible connection ∇; as in Chapter 2, this means that $d\langle s, s' \rangle = \langle \nabla s, s' \rangle + \langle s, \nabla s' \rangle$ for all $s, s' \in \Gamma(E)$. We form the twisted exterior derivative $d_\nabla : \Lambda^* T^* M \otimes E \to \Lambda^* T^* M \otimes E$ by the formula $d_\nabla(\omega \otimes s) = d\omega \otimes s + (-1)^{\deg(\omega)} \omega \wedge \nabla s$, where in the last term ω is wedged with the one-form part of ∇s. We let δ_∇ be the adjoint of d_∇ with respect to the inner product on $\Lambda^* \otimes E$ induced by

$$\langle \omega \otimes s, \eta \otimes s' \rangle = \int_M g(\omega, \eta) \langle s, s' \rangle \, \mathrm{dvol} \, .$$

Extend τ to an endomorphism on $\Lambda^* \otimes E$ by $\tau(\omega \otimes s) = \tau\omega \otimes s$. We define $\Lambda^{\pm}_E$ as before to be the ± 1 eigenspaces of τ, and finally define the *twisted signature* *operator*

$$D_\nabla = d_\nabla + \delta_\nabla : \Lambda_E^+ \to \Lambda_E^-.$$

While the index of D_∇ no longer has a topological interpretation, its multiplicativity properties suggest that we should have

$$\text{index}(D_\nabla) = \int_{M^{2k}} (L_k(\Omega_g)P(\Omega_\nabla))_{2k},$$

for some characteristic form P applied to the curvature Ω_∇ of ∇.

Exercise 21: *Define the* Chern character $\text{ch}(E) = [\text{ch}(\Omega_\nabla)] \in \bigoplus_k H_{dR}^{2k}(M)$ *of* E *by*

$$\text{ch} \begin{pmatrix} \lambda_1 & & \\ & \ddots & \\ & & \lambda_n \end{pmatrix} = \sum_{i=0}^{n} e^{\lambda_i}.$$

Show that $\text{ch}(E \oplus F) = \text{ch}(E) + \text{ch}(F)$ *for vector bundles* E, F. *Hint: Given connections* ∇_E, ∇_F *on* E, F, *define a sum connection on* $E \oplus F$ *whose curvature equals*

$$\begin{pmatrix} \Omega_{\nabla_E} & 0 \\ 0 & \Omega_{\nabla_F} \end{pmatrix}.$$

The methods of the last section lead to the following result.

Theorem 4.44 $\text{index}(D_\nabla) = \displaystyle\int_{M^{2k}} (L_k(\Omega_g)\text{ch}(\Omega_\nabla))_{2k}.$

We now assume familiarity with complex geometry. Let M be a complex Hermitian manifold of complex dimension n, and E a holomorphic Hermitian bundle over M with a compatible connection of type $(1,1)$. Let $\mathcal{O}(E)$ denote the sheaf of holomorphic sections of E, and define the arithmetic genus to be

$$\chi(M; E) = \sum_q (-1)^q \dim H^q(M, \mathcal{O}(E)),$$

where the right hand side denotes the sheaf cohomology groups of E. The Dolbeault theorem (the complex analogue of de Rham's theorem) and the Hodge theorem yield $\chi(M; E) = \text{index}(\bar{\partial}_E + \bar{\partial}_E^*)$, where $\bar{\partial}_E$ is the usual holomorphic exterior derivative $\bar{\partial}$ coupled to the connection on E as with the twisted signature operator, and

$$\bar{\partial}_E + \bar{\partial}_E^* : \bigoplus_k L^2 \Lambda^{0,2k}(M, E) \to \bigoplus_k L^2 \Lambda^{0,2k+1}(M, E)$$

acts on the L^2 spaces of E valued $(0, 2k)$-forms. Finally, define the *Todd polynomial* by the formal power series

$$\text{Td} \begin{pmatrix} \lambda_1 & & \\ & \ddots & \\ & & \lambda_{2n} \end{pmatrix} = \prod_{i=0}^{2n} \frac{\lambda_i}{1 - e^{-\lambda_i}}.$$

Theorem 4.45 (Hirzebruch-Riemann-Roch theorem)

$$\chi(M;E) = \int_M (\mathrm{Td}(\Omega_g)\mathrm{ch}(\Omega_\nabla))_{2n}.$$

Here g is the Hermitian structure on the complexified tangent bundle of M. Thus $\mathrm{Td}(\Omega_g)$ is a polynomial in the Chern classes of (the complexified tangent bundle of) M. It should be pointed out that the heat equation method has only been shown for Kähler manifolds, while the theorem holds for all complex manifolds. Again, the appearance of the Todd polynomial is motivated by the multiplicativity of the arithmetic genus for product manifolds and the fact that when E is trivial, $1 = \chi(\mathbf{CP}^n; E) = \int_{\mathbf{CP}^n} \mathrm{Td}(\Omega_g)$, where g is the Fubini-Study metric. When $\dim_{\mathbf{C}} M = 1$, this theorem reduces to the classical Riemann-Roch formula, from around 1870, and the formula for higher dimensions was more or less conjectured by Todd in the 1930s. A very nice historical account is in [20]. Thus Hirzebruch's theorem was a major advance for complex geometry.

The final example of an index theorem for a geometrically defined operator is given by the Dirac operator on a spin manifold. We will not define this operator, but just refer the reader to the thorough discussions in [5], [30], [59].

Roughly speaking, there is an index theorem for any operator which has a good Hodge theory. More precisely, from §1.3.4, if $D : \Gamma(E) \to \Gamma(F)$ is a differential operator between Hermitian vector bundles having Gårding's inequality for the "Laplacians" D^*D, DD^*, then the spaces of L^2 and C^∞ sections of E, F have a good decomposition with respect to D^*D, DD^*, respectively. The existence of such an inequality depends only on the top order part of D, as a look at the proof of Gårding's inequality in $\mathbf{R}^n$ reveals.

We now define the class of elliptic operators, which for our purposes are operators with Gårding's inequality. Given a differential operator $D: \Gamma(E) \to \Gamma(F)$ of order m acting on sections of bundles over M, and a point $x \in M$, choose a neighborhood of X over which both E and F are trivial. Over this neighborhood, D can be written as a matrix of partial differential operators:

$$D = \left(\sum_{I,|I| \leq m} a_{ij}^I(x)D^I \right)$$

for multi-indices $I = (i_1, \ldots, i_k)$ in the notation of (1.16). Now form the top order *symbol matrix* by replacing each D^I of order m by the scalar $\xi_1^{i_1} \cdot \ldots \cdot \xi_n^{i_n}$ for $\xi = (\xi_1, \ldots, \xi_n) \in \mathbf{R}^n$. (It is a theorem that this matrix is independent of trivializations chosen if ξ is considered as an element of $T_x^* M$.) The resulting matrix is

$$\sigma(D)(x,\xi) = \left(\sum_{I,|I|=m} a_{ij}^I(x)\xi_1^{i_1} \cdot \ldots \cdot \xi_n^{i_n} \right).$$

Definition: D is *elliptic* if $\sigma(D)(x,\xi)$ is invertible for every $x \in M$ and for every nonzero $\xi \in T_x^* M$.

Note that the symbol is determined by its values on the unit sphere bundle in $T_x^* M$.

Exercise 22: *(i) Verify that $d + \delta$ taking even forms to odd forms is an elliptic operator. Hint: use Lemma 2.39 to show that*

$$\sigma(d + \delta)(x, \xi)(\theta^J) = \xi \wedge \theta^J + \iota_\xi \theta^J,$$

where $\{\theta^i\}$ is an orthonormal frame of $T_x^* M$ and ι_ξ denotes interior product. Show also that the Laplacian on k-forms, Δ^k, is an elliptic operator, with symbol

$$\sigma(\Delta^k)(x, \xi)(\omega) = g^{ij}(x)\xi_i\xi_j\omega = |\xi|^2\omega.$$

Note that $g^{ij}(x)\xi_i\xi_j$ is well defined only if ξ is a cotangent vector. Hint: Use Exercise 33, Chapter 1, to reduce the problem to the case $k = 0$. Then use (1.14).

(ii) Let E, F be trivial line bundles over a manifold of dimension two. Let D be a positive definite second order differential operator with real coefficients. Show that D is elliptic iff the curve $\sigma(D)(x, \xi) = 1$ is an ellipse in $T_x^* M$ for all $x \in M$.

For the elliptic operator $d + \delta$ on even forms, the symbol is a universal polynomial in the vector ξ valid on any Riemannian manifold. For general elliptic operators, the symbol will not have such a universal nature, and so we expect the index to depend on the unit sphere bundle $\Sigma M \subset T^* M$. Thus if the index is a topological invariant, we expect this invariant to be given by characteristic class information on TM, not just on M.

The Atiyah-Singer index theorem identifies the index of an elliptic operator over M with the integral over TM of a characteristic class on TM. This characteristic class has two components: (i) the Chern character of the symbol bundle $\sigma = \sigma(D, E, F)$, a vector bundle associated to the data (D, E, F), and (ii) the Todd class $\mathrm{Td}(M)$ of $TM \otimes C$, the complexified tangent bundle (pulled back to TM).

To define the symbol bundle, we follow [62]. Let S^+, S^- denote the upper and lower hemispheres of S^n respectively, for $n = \dim M$. Notice that S^n is the two point suspension of the equator $S^+ \cap S^-$. Glue $S^- \times E_x$ to $S^+ \times F_x$ along the equator by identifying (ξ, v) with $(\xi, \sigma(D)(x, \xi)(v))$, where $S^+ \cap S^-$ is identified with the unit cotangent vectors $\Sigma_x M$. This gives us a vector bundle over S^n for each x, and as x varies we obtain a vector bundle Σ over $S^n M$, the suspension bundle associated to ΣM.

We have to relate Σ to a bundle over TM. We can take sums of vector bundles with respect to Whitney sum, so formally we can take differences of bundles. (More precisely, we pass from the semigroup of vector bundles over M to the group of virtual bundles $K(M)$, which is the free group with generators given by isomorphism classes of vector bundles over M modulo the relation $E_1 \oplus E_2 - (E_1 \oplus E_2)$.) The difference $\Sigma - F$ is trivial over S^+, and so defines a virtual bundle $\Sigma - F$ over $S^n M / S^+ M$. Let $B^n M$ denote the unit disk bundle in

T^*M. Then $S^n M/S^+M$ is diffeomorphic to $B^n M/\Sigma M$, so we define the symbol bundle $\sigma(D, E, F)$ over $B^n M/\Sigma M$ to be $\Sigma - F$. The Chern character is linear with respect to Whitney sum, and so extends to a homomorphism

$$\text{ch}: K(S^n M/S^+M) \to H^*(S^n M/S^+M; \mathbf{Q}) \cong H^*(B^n M, \Sigma M; \mathbf{Q}).$$

In particular,

$$\text{ch}(\sigma(D, E, F)) \in H^*(B^n M, \Sigma M; \mathbf{Q}) \cong H_c^*(TM; \mathbf{Q}),$$

where H_c^* denotes cohomology with compact support. We can pick a compactly supported form on TM representing $\text{ch}(\sigma(D, E, F))$ and another form representing the Todd class. We denote these forms just by the cohomology class they represent.

Assume M is oriented. If $\{X_1, \ldots, X_n\}$ is a positively oriented local frame on M, we orient TM so that $\{X_1, X_1', \ldots, X_n, X_n'\}$ is a positively oriented frame, where the prime indicates belonging to the second factor in the natural decomposition $TTM \cong TM \oplus TM$. Note that since $\text{ch}(\sigma(D, E, F))$ can be represented by a compactly supported form, we can integrate $\text{ch}(\sigma(D, E, F)) \wedge \omega$ over TM for any form ω on TM.

Theorem 4.46 (Atiyah-Singer Index Theorem) *Let E, F be complex vector bundles over a compact manifold M, and let $D : \Gamma(E) \to \Gamma(F)$ be an elliptic differential operator. Then*

$$\text{index}(D) = (-1)^n \int_{TM} \text{ch}(\sigma(D, E, F))\text{Td}(M).$$

For geometric operators, the right hand side of the index theorem can be rewritten as the expected integral over M. Thus we can trace a direct line from the Gauss-Bonnet theorem in 1850 to the Atiyah-Singer theorem of 1963. In fact, the index theorem can be reduced to the case of the twisted signature operators by standard results in the K-theory of vector bundles [30], so in this sense a heat equation proof of the full index theorem is known.

Chapter 5

Zeta Functions of Laplacians

In this chapter we will encode the spectral information of a Laplacian-type operator into a zeta function first introduced by Minakshisundaram and Pleijel [48] and Seeley [61]. While this is theoretically equivalent to the encoding of the spectrum given by the trace of the heat operator, the zeta function contains spectral information hard to obtain by heat equation methods. In particular, the important notion of the determinant of a Laplacian is given in terms of the zeta function.

In §5.1, we introduce the zeta function and use it to produce new conformal invariants in Riemannian geometry. In §5.2, we outline Sunada's elegant construction of isospectral, nonhomeomorphic four-manifolds. While the results in §5.1 are conceivably obtainable directly from the heat operator, the results in §5.2 depend on the zeta function for motivation. Finally, in §5.3 we discuss the determinants of Laplacians on forms and define analytic torsion, which we show is a smooth invariant subtler than the invariants produced in Chapter 4. We conclude with an overview of recent work of Bismut and Lott connecting analytic torsion with Atiyah-Singer index theory for families of elliptic operators. This last discussion is the most difficult part of the book and contains no proofs.

5.1 The Zeta Function of a Laplacian

By a Laplacian-type operator, we mean any symmetric second order elliptic differential operator $\Delta : \Gamma(E) \to \Gamma(E)$ acting on sections f of a Hermitian bundle E over a compact n-manifold M satisfying $\langle \Delta f, f \rangle \geq C \langle f, f \rangle$ for some $C \in \mathbf{R}$. The basic examples are the Laplacians on forms, where $C = 0$. For the application below, note that since the definition of ellipticity depends only on the top order part of the operator, changing lower order terms does not affect the ellipticity.

As mentioned in Chapter 4, these elliptic operators have smooth heat kernels with corresponding heat operators satisfying $(\partial_t + \Delta)(e^{-t\Delta}\omega) = 0$ for $\omega \in \Gamma(E)$. By the appropriate Gårding inequality, the space $L^2(E)$ of L^2 sections of E decomposes into finite dimensional eigenspaces for Δ with eigenvalues, say, $\{\lambda_i\}$. The condition $\langle \Delta f, f \rangle \geq C\langle f, f \rangle$ guarantees that the number of negative eigenvalues is finite. For $s \in \mathbf{C}$, we now set the zeta function of Δ to be

$$\zeta(s) = \sum_{\lambda_i \neq 0} \lambda_i^{-s}. \tag{5.1}$$

(If some of the λ_i are negative, replace λ_i^{-s} by $(\mathrm{sgn}\,\lambda_i)|\lambda_i|^{-s}$.) For example, if $\Delta = \Delta^0$ on S^1, then it should be easy to check that

$$\zeta(s) = \zeta_R(2s),$$

where $\zeta_R(s)$ is the Riemann zeta function.

We now show that the zeta function exists as a meromorphic function on $\mathbf{C}$. Let the heat kernel for Δ have the asymptotic expansion

$$e(t, x, x) \sim \sum_{k=0}^{\infty} \left(\int_M u_k(x, x) \, \mathrm{dvol} \right) t^{k-(n/2)}.$$

(We have absorbed the factors of $(4\pi)^{-n/2}$ into the u_k.) The existence of the heat kernel and its asymptotic expansion is covered in [30]. Denote the greatest integer function by $[\,\cdot\,]$.

Theorem 5.2 $\zeta(s)$ *converges and is holomorphic for* $\mathrm{Re}(s) > \frac{n}{2}$, *and* $\zeta(s)$ *has meromorphic continuation to* $\mathbf{C}$ *with at worst simple poles, occurring only at* $s = \frac{n}{2}, \frac{n}{2} - 1, \frac{n}{2} - 2, \ldots, \frac{n}{2} - [\frac{n-1}{2}]$. *In particular,*

$$\zeta(0) = \begin{cases} -\dim \mathrm{Ker}\,\Delta, & \dim M \text{ odd}, \\ \int_M u_{n/2} - \dim \mathrm{Ker}\,\Delta, & \dim M \text{ even}. \end{cases}$$

PROOF. Since $\{\lambda_i : \lambda_i < 0\}$ is a finite set,

$$\sum_{\{i : \lambda_i < 0\}} (\mathrm{sgn}\,\lambda_i)|\lambda_i|^{-s}$$

exists for all $s \in \mathbf{C}$. Thus we are reduced to the case where the λ_i are all nonnegative. We introduce the *Mellin transform* of the zeta function. Since

$$\Gamma(s) = \int_0^\infty t^{s-1} e^{-t} dt,$$

a change of variables gives

$$\lambda_i^{-s} = \frac{1}{\Gamma(s)} \int_0^\infty t^{s-1} e^{-t\lambda_i} dt.$$

Thus, wherever the zeta function converges, we have

$$
\begin{aligned}
\zeta(s) &= \sum_{\lambda_i \neq 0} \lambda_i^{-s} = \frac{1}{\Gamma(s)} \int_0^\infty t^{s-1} \left(\sum_i e^{-t\lambda_i} - \dim \operatorname{Ker} \Delta \right) dt \\
&= \frac{1}{\Gamma(s)} \int_0^\infty t^{s-1} \operatorname{Tr}\left(e^{-t\Delta} - P\right) dt \\
&= \frac{1}{\Gamma(s)} \int_0^1 t^{s-1} \operatorname{Tr}(e^{-t\Delta} - P)\, dt + \frac{1}{\Gamma(s)} \int_1^\infty t^{s-1} \operatorname{Tr}(e^{-t\Delta} - P)\, dt \\
&= \text{(a)} + \text{(b)}, \tag{5.3}
\end{aligned}
$$

say, where P is the orthogonal projection onto Ker Δ. As the reader should check, (b) is of the form

$$
\frac{1}{\Gamma(s)} \int_1^\infty t^{s-1} O(e^{-\tilde{\lambda}t})\, dt,
$$

where $\tilde{\lambda}$ is the first nonzero eigenvalue of Δ. Since the gamma function is never zero, (b) is a holomorphic function of s. Moreover, for fixed $N > \frac{n}{2}$,

$$
\begin{aligned}
\text{(a)} &= \frac{1}{\Gamma(s)} \int_0^1 t^{s-1} \left(\sum_{k=0}^N \int u_k t^{k-\frac{n}{2}} + O(t^{N+1-\frac{n}{2}}) - \dim \operatorname{Ker} \Delta \right) dt \\
&= \frac{1}{\Gamma(s)} \left(\sum_{k=0}^N \frac{1}{s+k-\frac{n}{2}} \int u_k - \frac{\dim \operatorname{Ker} \Delta}{s} + R(s) \right),
\end{aligned}
$$

where $R(s)$ is some bounded function coming from the integration of the term $O(t^{N+s-\frac{n}{2}})$. Thus (a) converges for all $s \in \mathbf{C}$ with $\operatorname{Re}(s) > \frac{n}{2}$ and has a meromorphic continuation to all of $\mathbf{C}$ with simple poles occurring only at the points stated in the theorem. Finally, because $\Gamma(s)$ has a simple pole at $s = 0$, the term (b) is zero at $s = 0$, while (a) gives $\int_M u_{n/2} - \dim \operatorname{Ker} \Delta$ if dim M is even and $-\dim \operatorname{Ker} \Delta$ if dim M is odd.

Exercise 1: *Calculate the special values $\zeta((n/2) - k)$ (or the residues at the poles) for $k \in \{0, 1, 2, \ldots\}$. Hint: the answer should be multiples of the $\int_M u_k$.*

Remarks: (1) Note the curious result that the nonzero spectrum of the Laplacian determines the multiplicity of the zero spectrum in odd dimensions.

(2) The inverse Fourier transform takes $\sum e^{-\lambda_i t}$, the trace of the heat operator, to a sum of delta functions, concentrated at the values of the λ_i, with mass the multiplicity of each eigenvalue. Thus the spectrum and the trace of the heat kernel determine each other. The Mellin transform and its inverse show that in turn the trace of the heat kernel and the zeta function determine each other. Since the zeta function is meromorphic with simple poles, the value of the zeta function on any open set in $\mathbf{C}$ thus determines the spectrum of Δ. Of course,

in practice this means that we don't stand a chance of explicitly computing the zeta function on any open set.

(3) Let $\zeta_k(s)$ denote the zeta function for the Laplacian on k-forms. As in Theorem 4.4, the Chern-Gauss-Bonnet theorem is equivalent to the statement

$$\sum_k (-1)^k \zeta_k(0) = 0$$

together with the identification of $\sum (-1)^k u_{n/2}^k$ with the Euler form. In fact, a proof of the Atiyah-Singer index theorem can be framed in terms of zeta functions, by considering the zeta functions for the Laplacians $D_\nabla^* D_\nabla, D_\nabla D_\nabla^*$ for twisted signature operators. Of course, the crucial identification of the heat kernel asymptotics with the appropriate characteristic form proceeds as before.

As an application, we use zeta function methods to produce conformal invariants of a Riemannian metric. Historically, several tensors (e.g. the Weyl and Schouten tensors) were produced whose pointwise values were independent of conformal change in the metric. There is still no systematic method for writing down all such tensors, and new examples seem hard to come by. The invariants we produce are global ones, but they have pointwise analogues, as explained at the end of the section.

We define the *conformal Laplacian* $\Box : L^2(M) \longrightarrow L^2(M)$ by

$$\Box = \Box_g = \Delta + \frac{n-2}{4(n-1)}s,$$

where s is the scalar curvature. Note that in dimension two, the conformal and ordinary Laplacians coincide. Recall that, for $f \in C^\infty(M)$, the change of metric $g \mapsto e^{2f}g$ is conformal, i.e. angles between intersecting curves are unchanged. While the ordinary Laplacian changes rather badly under a conformal change of metric, the conformal Laplacian has a simple transformation law.

Exercise 2: *Show that*

$$\Box_{e^{2f}g} = e^{-(\frac{n}{2}+1)f}\Box_g e^{(\frac{n}{2}-1)f}. \tag{5.4}$$

Conclude that $\dim \operatorname{Ker} \Box_g = \dim \operatorname{Ker} \Box_{e^{2f}g}$, *and that the variation of the conformal Laplacian is given by*

$$\delta_f \Box_g \equiv \left.\frac{d}{dp}\right|_{p=0} \Box_{e^{2pf}g} = -2f\Box_g.$$

With the same notation, show that $\delta_f e^{-t\Box_g} = -t(\delta_f \Box_g)e^{-t\Box_g}$. *Hint: (5.4) can be checked directly for* $n = 2$. *For* $n \neq 2$, *first show that the scalar curvature* s_f *for* $g_1 = e^{2f}g$ *satisfies*

$$s_f = \frac{4(n-1)}{n-2}e^{-(n+2)f/2}\Box_g e^{(n-2)f/2}$$

(see [3, Ch. 6]). Conclude that

$$\Box_{g_1} 1 = e^{-(n+2)f/2} \Box_g e^{(n-2)f/2}.$$

Pick an arbitrary function h on M and set $g_2 = e^{2h} g_1 = e^{2h+2f} g$. Now express $\Box_{g_2} 1$ in terms of $\Box_g, \Box_{g_1}$ as above to conclude that

$$\Box_{g_1} e^{(n-2)h/2} = e^{-(n+2)f/2} \Box_g e^{-(n-2)f/2} e^{-(n-2)h/2}.$$

This proves the exercise for positive functions $e^{-(n-2)h/2}$, and the rest of the exercise follows.

If the two exponents $\frac{n}{2} \pm 1$ in the last equation were both, say, $\frac{n}{2} + 1$, then $\Box_{e^{2f} g}$ would be conjugate to $\Box_g$, and the two operators would be isospectral. The next theorem shows that the special value $\zeta(0)$ is unchanged, and so provides a conformal invariant of M. Moreover, in odd dimensions, where the conformal invariance of $\zeta(0)$ is trivial, $\zeta'(0)$ is a conformal invariant if Ker $\Box_g$ vanishes. The significance of $\zeta'(0)$ will be discussed in §5.3.2.

Exercise 3: *Show that $\zeta'(0)$ cannot be a conformal invariant if $\dim \text{Ker } \Box_g \neq 0$ by computing the change in $\zeta'(0)$ under a scaling of the metric $g \mapsto \lambda^2 g$ for some $\lambda \in \mathbf{R}$.*

Theorem 5.5 *(i)* $\zeta_{\Box_g}(0) = \zeta_{\Box_{e^{2f}g}}(0)$.
     (ii) If $\dim M$ is odd and Ker $\Box_g = 0$, then $\zeta'_{\Box_g}(0) = \zeta'_{\Box_{e^{2f}g}}(0)$.

PROOF. (i) The result is trivial in odd dimensions by the last theorem and Exercise 2. In even dimensions, fix g and f. Assume that $\Box_g$ is a positive operator (as happens, e.g., if the scalar curvature is positive.) Thus the heat operator $e^{-t\Box_{\exp(2pf)g}}$ has exponential decay as $t \to \infty$ for all p close to zero. In the notation of Exercise 2, we have

$$
\begin{aligned}
\delta_f \zeta_{\Box_g}(0) &= \delta_f \frac{1}{\Gamma(s)} \int_0^\infty t^{s-1} \text{Tr}\left(e^{-t\Box_g}\right) dt \Big|_{s=0} \\
&= \frac{1}{\Gamma(s)} \int_0^\infty t^{s-1} \delta_f \text{Tr}\left(e^{-t\Box_g}\right) dt \Big|_{s=0} \\
&= \frac{1}{\Gamma(s)} \int_0^\infty t^{s-1} \text{Tr}\left(-t\delta_f \Box_g e^{-t\Box_g}\right) dt \Big|_{s=0} \\
&= -\frac{2}{\Gamma(s)} \int_0^\infty t^s \text{Tr}\left(f\Box_g e^{-t\Box_g}\right) dt \Big|_{s=0} \\
&= \frac{2}{\Gamma(s)} \int_0^\infty t^s \partial_t \text{Tr}\left(f e^{-t\Box_g}\right) dt \Big|_{s=0} \\
&= -\frac{2}{\Gamma(s)} \int_0^\infty s t^{s-1} \text{Tr}\left(f e^{-t\Box_g}\right) dt \Big|_{s=0}.
\end{aligned}
\tag{5.6}
$$

The reader should check that the interchange of δ_f and the integral is valid. There are no boundary terms in the integration by parts, as the integrand is of exponential decay at infinity, and an examination of the asymptotic expansion near zero shows that for $\text{Re}(s)$ sufficiently large, the integrand also vanishes at zero.

Exercise 4: *Let the heat kernel for $\square_g$ have the asymptotic expansion*

$$e(t, x, x) \sim \sum_{k=0}^{\infty} u_k(x) t^{k-(n/2)}.$$

Show that

$$\text{Tr } f e^{-t\square_g} \sim \sum_{k=0}^{\infty} \left(\int_M f(x) u_k(x) \, \text{dvol} \right) t^{k-(n/2)}.$$

Hint: What is the kernel of $\text{Tr } f e^{-t\square_g}$?

Returning to the proof of the theorem, we have

$$\delta\zeta_{\square_f}(0) = -\frac{2s}{\Gamma(s)} \left(\int_0^1 t^{s-1} \text{Tr} \left(f e^{-t\square_g} \right) dt + \int_1^\infty t^{s-1} \text{Tr} \left(f e^{-t\square_g} \right) dt \right) \Big|_{s=0}$$

$$= -\frac{2s}{\Gamma(s)} \left(\sum_{k=0}^{N} \frac{\int_M f u_k}{s+k-n/2} + R(s) + \int_1^\infty t^{s-1} \text{Tr} \left(f e^{-t\square_g} \right) dt \right) \Big|_{s=0},$$

in the notation of the proof of Theorem 5.2. Now in the last expression $2s/\Gamma(s)$ has a double zero at $s = 0$, while the sum over k has a simple pole for $k = \frac{n}{2}$ and the last integral is holomorphic in s. Thus the last expression vanishes at $s = 0$, which proves the theorem for positive conformal Laplacians.

In general, the scalar curvature term $(n-2)/(4(n-1)) \cdot s$ is bounded below by a constant $C = C(g)$. Therefore, $\langle \square_g h, h \rangle \geq \langle \Delta h, h \rangle + C \langle h, h \rangle \geq C \langle h, h \rangle$, so $\square_g$ can have only a finite number of negative eigenvalues. In such a case, expressions such as

$$\int_0^\infty t^{s-1} \text{Tr}(e^{-t\square_g}) \, dt$$

diverge for all s. However, the finite contribution to the zeta function coming from these nonpositive eigenvalues can be handled directly as in Theorem 5.2. The infinite number of positive eigenvalues are then treated as above. Details are in [54].

The proof of (ii) is similar. Near $s = 0$ we have

$$\delta_f[\Gamma(s)\zeta(s)] = \Gamma(s)[\delta_f\zeta(0) + s\delta_f\zeta'(0) + \text{O}(s^2)].$$

Since $\delta_f\zeta(0) = 0$ and $\lim_{s\to0} s\Gamma(s) = 1$, this implies

$$\delta_f\zeta'(0) = \delta_f[\Gamma(s)\zeta(s)]|_{s=0} = \delta_f \int_0^\infty t^{s-1} \text{Tr}(e^{-t\square_g}) \, dt|_{s=0}. \qquad (5.7)$$

Proceeding as in (5.6), we obtain

$$\delta_f \zeta'(0) = -2s \left(\sum_{k=0}^{N} \frac{\int_M f u_k}{s+k-n/2} + R(s) + \int_1^\infty t^{s-1} \text{Tr}\left(f e^{-t\square_g}\right) dt \right) \Bigg|_{s=0}.$$

$$(5.8)$$

Since n is odd, the expression inside the large parentheses has no pole at $s = 0$, so (5.8) vanishes at $s = 0$. (If Ker $\square_g \neq 0$, then that expression does contain a term of the form A/s, so the argument breaks down; again details are in [54].)

Thus $\zeta(0)$ is a conformal invariant for the conformal Laplacian. Calculations show that it is highly nontrivial. For example, when dim $M = 6$, we get

$$\zeta(0) = -\frac{1}{9 \cdot 7!} \left(30\chi(M) - (4\pi)^{-3} \int_M (54\Omega_6 + 204B_1 + 47B_2) \right),$$

where Ω_6 is the Fefferman-Graham conformal invariant defined by

$$\Omega_6 = |\nabla W|^2 + 2 \left(4B_1 + B_2 - \frac{1}{5}s|W|^2 - \langle W, \nabla^i \nabla_i W \rangle \right.$$
$$\left. - \frac{1}{2}(\nabla_k B_{ij} \nabla^i B_{kj}) - |\nabla B|^2 \right) - \frac{1}{45}|\nabla s|^2.$$

Here $W = W^i_{jkl}$ is the Weyl tensor, a pointwise conformal invariant given by

$$W^i_{jkl} = R^i_{jkl} - \frac{1}{n-2}(R_{jk}\delta^i_l - R_{jl}\delta^i_k + g_{jk}R^i_l - g_{jl}R^i_k)$$
$$+ \frac{s}{(n-1)(n-2)}(g_{jk}\delta^i_l - g_{jl}\delta^i_k),$$

$\nabla^i = g^{ij}\nabla_j$, $B = B_{ij} = R_{ij} - \frac{s}{n}g_{ij}$ is the traceless Ricci tensor built from the Ricci and scalar curvatures, $B_1 = W^i{}_j{}^k{}_l W_i{}^p{}_k{}^q W^j{}_p{}^l{}_q$, and $B_2 = W^{ij}{}_{kl} W^{kl}{}_{pq} W^{pq}{}_{ij}$. The length of all tensors is computed with respect to the Riemannian metric. As explained in §5.3, there is at present no method for computing $\zeta'(0)$ except in special cases.

It also turns out that in even dimensions, $u_{(n/2)-1}(x)$ transforms very simply under a conformal change of the metric, for each $x \in M$. If we define a pointwise zeta $\zeta(s, x)$ by

$$\zeta(s, x) = \sum_{\lambda_i \neq 0} e^{-\lambda_i t} |\phi_i(x)|^2,$$

where the ϕ_i are the eigenfunctions for λ_i, then as above

$$\text{res}_{s=1}\zeta(s, x) = u_{(n/2)-1}(x).$$

Thus the theory of spectral zeta functions produces pointwise conformal invariants $\text{res}_{s=1}\zeta(s, x)$, global conformal invariants $\zeta(0)$ given by integrating local expressions, and global nonlocal invariants $\zeta'(0)$.

5.2   Isospectral Manifolds

Recall that isospectral manifolds have many geometric quantites in common, such as their volume and the integral of the scalar curvature. Around 1960, M. Kac raised the question of whether isospectral manifolds were in fact isometric. Many counterexamples have been given in the past 35 years. In this section, we describe a zeta function approach to producing isospectral but nonisometric manifolds. The key idea is that two manifolds are isospectral iff their zeta functions are identical, by the remarks in §5.1.

It is well known that there is a strong analogy between finite Galois extensions of a field and finite coverings of a manifold. In a beautiful paper [65], Sunada extended this analogy to zeta functions. He first notes the following result from field theory.

Proposition 5.9 *Let K be a finite Galois extension of* **Q** *with Galois group G, and let K_1, K_2 be subfields of K corresponding to subgroups H_1, H_2 of G, respectively. Then the following are equivalent:*
(i) Each conjugacy class of G meets H_1 in the same number of elements as it meets H_2.
(ii) The Dedekind zeta functions of K_1, K_2 satisfy $\zeta_{K_1}(s) = \zeta_{K_2}(s)$.

Recall that the conjugacy class of $g \in G$ is $[g] = \{sgs^{-1} : s \in G\}$. The collection of conjugacy classes of elements of G forms a partition of G.

Sunada carries the proof of this theorem, which depends on a simple trace formula, over to the case of manifolds.

Theorem 5.10   (Sunada's Theorem) *Let M be a finite covering of a compact manifold M_0 with deck transformation group G. Let M_1 and M_2 be covers of M_0 corresponding to subgroups $H_1, H_2 \subset G$. If H_1, H_2 have the property (i) of the previous proposition, then $\zeta_{M_1}(s) = \zeta_{M_2}(s)$ with respect to the metrics pulled back from any metric on M_0. In particular, M_1 is isospectral to M_2.*

Note that if H_1, H_2 are nonisomorphic, then M_1, M_2 are not even homeomorphic. Although examples of isospectral, nonhomeomorphic manifolds had been given previously [35], Sunada's theorem gives the first systematic method of producing such examples.

For example, if H_1, H_2 are any finite groups of the same order c, with exponent p (i.e. $x^p = 1$ for all $x \in H_1, H_2$) for some prime $p \neq 2$, then H_1, H_2 will have condition (i) of the proposition, when considered as subgroups of the group G of permutations on c letters by Cayley's theorem.

To show this, let $[x]$ denote the conjugacy class of $x \in G$. If $x = 1$, condition (i) is clear. If $[x] \cap (H_1 \cup H_2) = \emptyset$, condition (i) again trivially holds. Since $[x] = [y]$ if x and y are conjugate, we are reduced to the case $x \in (H_1 \cup H_2) \setminus \{1\}$. Now x has prime order p, so as a permutation of $\{1, \ldots, c\}$, x determines a partition of $\{1, \ldots, c\}$ into disjoint sets $A_1, \ldots, A_{c/p}$ of order p on each of which x acts by cyclic permutation. Pick $y \in (H_1 \cup H_2) \setminus \{1\}$ with associated partition $B_1, \ldots, B_{c/p}$. Choose $a_i \in A_i, b_i \in B_i$ arbitrarily. Call g the permutation which sends a_i^k to b_i^k for $k = 1, \ldots, p$. Then it is easy to check that $g^{-1}yg = x$. Thus all $x \in (H_1 \cup H_2) \setminus \{1\}$ are conjugate. For such an x, $[x] \cap H_i = H_i$, and so condition (i) is satisfied.

In particular, an example is furnished by $H_1 = (\mathbf{Z}_p)^3$ and H_2 the group with generators a, b and relations $a^p = b^p = [a, b]^p = 1$, $a[a, b] = [a, b]a$, $b[a, b] = [a, b]b$. Since H_1 is abelian and H_2 is not, these groups are nonisomorphic. A concrete realization of H_2 is given by the set of 3×3 upper triangular matrices with coefficients in $\mathbf{Z}_p$ and with diagonal entries one.

It is a standard result in topology that we can construct a compact four-manifold with an arbitrary finitely presented fundamental group. Therefore, if we construct M_0 with fundamental group G, we have covering spaces M_1, M_2 corresponding to H_1, H_2 which are isospectral but nonhomeomorphic.

We now begin the proof of Sunada's theorem. The discussion is based on the fuller treatment in [13, Ch. 11]. Let a finite group G act on a finite dimensional vector space V. Thus each $g \in G$ gives a linear transformation $g : V \to V$, and for $v \in V$ we will just write gv for $g(v)$. Note that the trace of g as a linear transformation depends only on the conjugacy class of g. For a subgroup $H \subset G$, we set V^H to be the set of H invariant vectors: $V^H = \{v \in V : hv = v, \forall h \in H\}$. Finally, for $h \in H$ we denote the conjugacy class of h within H by $[h]' = \{shs^{-1} : s \in H\}$.

Lemma 5.11 *We have*

$$\dim(V^H) = \frac{1}{\#H} \sum_{[h]'} \#[h]' \cdot \operatorname{tr}(h) = \frac{1}{\#H} \sum_{[g]} \#([g] \cap H) \cdot \operatorname{tr}(g).$$

PROOF. The projection $P : V \to V$ given by

$$Pv = \frac{1}{\#H} \sum_{h \in H} hv$$

clearly maps V to V^H and is the identity on V^H. Computing the trace of P on a basis of V^H extended to a basis of V yields $\operatorname{tr}(P) = \dim(V^H)$. Thus

$$\dim(V^H) = \frac{1}{\#H} \sum_{h \in H} \operatorname{tr}(h) = \frac{1}{\#H} \sum_{[h]'} \#[h]' \cdot \operatorname{tr}(h),$$

which proves the first equation.

For $h \in H$, we certainly have $[h]' \subset [h]$. Thus for $g \in G$, we have either $[h]' \subset [g]$ or $[h]' \cap [g] = \emptyset$. We have

$$\sum_{[h]'} \#[h]' \cdot \operatorname{tr}(h) = \sum_{[g]} \sum_{[h]' \subset [g]} \#[h]' \cdot \operatorname{tr}(g) = \sum_{[g]} \#([g] \cap H) \cdot \operatorname{tr}(g),$$

as clearly $\sum_{[h]' \subset [g]} \#[h]' = \#([g] \cap H)$.

As a corollary, we see that if H_1, H_2 satisfy condition (i) of Proposition 5.9, then $\dim(V^{H_1}) = \dim(V^{H_2})$.

PROOF OF THEOREM 5.10: Let $E_\lambda(M_i)$ denote the (possibly empty) space of eigenfunctions of the Laplacian on M_i with eigenvalue λ. Let $u_1, \ldots, u_k$ be a basis of $E_\lambda(M_1)$ and let $v_1, \ldots, v_k$ be their lifts to M. The set $\{gv_i : g \in G, i = 1, \ldots, k\}$ spans a subspace V of $L^2(M)$ which is easily seen to be G-invariant. Since M has the pullback metric from M_0, we have $V \subset E_\lambda(M)$. Now there is a one-to-one correspondence between elements of $L^2(M)$ which are H_2 invariant and elements of $L^2(M_2)$. In particular, there is a one-to-one correspondence between $E_\lambda^{H_2}(M)$ and $E_\lambda(M_2)$. Thus there is an injection from V^{H_2} to $E_\lambda(M_2)$. The dimension of V^{H_2} is at least $k = \dim(E_\lambda(M_1))$, as $\sum_{h \in H_2} hv_i$ are linearly independent functions in V^{H_2}. Thus $\dim(E_\lambda(M_1)) \leq \dim(E_\lambda(M_2))$. Reversing the roles of M_1, M_2 shows that $\dim(E_\lambda(M_1)) = \dim(E_\lambda(M_2))$.

5.3 Reidemeister Torsion and Analytic Torsion

The Atiyah-Singer index theorem can be viewed as providing analytic/geometric interpretations of certain topological information contained in the cohomology ring. Of course, manifolds may contain more topological structure than can be detected by cohomology. In particular, there exist lens spaces (quotients of S^3 by finite group actions) which have the same cohomology rings and homotopy groups but which are nonhomeomorphic. The topological invariant used to distinguish these two spaces is the Reidemeister torsion, introduced by Reidemeister and Franz in the 1930s for this purpose. We call this invariant "precohomological," to emphasize that it is defined at the cochain level.

With the success of the index theorem in the 1960s, it was natural to look for an analytic expression for the Reidemeister torsion. Such an expression of course must be more refined than the index of an elliptic operator. In the early 1970s, Ray and Singer [58] defined the analytic torsion in analogy with the Reidemeister torsion, and showed that the analytic torsion possessed many of the key properties of Reidemeister torsion. Their definition of analytic torsion depends crucially on the zeta function for the Laplacians on forms, in contrast to the index theorem, for which the heat equation and zeta function methods are equivalent. Around 1980, Cheeger[15] and Müller [51] independently proved the equality of analytic and Reidemeister torsion; a more natural proof is due to Bismut and Zhang [9]. Since then, analytic torsion has figured prominently in questions in mathematical physics.

In §5.3.1, we define the torsion of a general complex and then specialize to the case of Reidemeister torsion. In §5.3.2, we define the analytic torsion, prove that it is a smooth invariant, and comment on the proof of the equality of the two torsions. In §5.3.3, we indicate a deep relationship between index theory and analytic torsion uncovered by Bismut and Lott [8].

5.3.1 Reidemeister Torsion

The torsion of an exact complex generalizes the volume of a linear transformation. If $T : C^1 \to C^2$ is a linear isomorphism of finite dimensional inner product spaces, then it makes no sense to talk about the determinant, or volume distortion, of T, as there is no natural choice of bases of C^1, C^2. As the next best option, we define the volume of T to be $\sqrt{\det T^*T}$, as this quantity measures the volume distortion of T if $C^1 = C^2$. More generally, given an exact complex of inner product spaces

$$0 \to C^1 \xrightarrow{d^1} C^2 \xrightarrow{d^2} \cdots \xrightarrow{d^{n-1}} C^n \to 0,$$

we can form "Laplacians" $\Delta^q = (d^q)^*d^q + d^{q-1}(d^{q-1})^* : C^q \to C^q$ and define the torsion $T = T(C^*, d^*)$ by

$$\log T = \frac{1}{2}\sum_q (-1)^{q+1}q \ \log\det \Delta^q. \tag{5.12}$$

We won't attempt to justify the particular linear combination chosen in (5.12), but just refer the reader to [58, §1].

Now let M be a compact manifold equipped with a triangulation $\{e_j\}$ and a unitary representation $\rho : \pi_1(M) \to U(N)$. We lift the triangulation to one of the universal cover $\tilde{M}$ and form the finite dimensional space of twisted k-chains $C_k(M;\rho)$; this space has as a basis $(\tilde{e}_j, v)$, with $\tilde{e}_j$ a lifted k-simplex and $v \in \mathbf{C}^N$, modulo the identification $(\tilde{e}_j, v) \sim (\gamma \cdot \tilde{e}_j, \rho(\gamma^{-1})v)$, for all $\gamma \in \pi_1(M)$. There is a natural boundary operator $\partial : C_k \to C_{k-1}$ induced by $\partial(\tilde{e}_j, v) = (\partial e_j, v)$, where ∂ is the boundary operator for the triangulation, and so the dual spaces $C^k(M;\rho)$ of k-cochains have a natural coboundary operator δ^k. We assume that the representation is acyclic, which means that the cochain complex (C^k, δ^k) is exact. Put an inner product on the equivalence classes of twisted cochains $C^k(M;\rho)$ by setting

$$\langle [\tilde{e}_j, v], [\tilde{e}_k, w]\rangle = \langle \rho(\gamma^{-1})(v), w\rangle_{\mathbf{C}^N}$$

if there exists γ such that $\gamma \cdot \tilde{e}_j = \tilde{e}_k$, and 0 otherwise. This gives all the data needed to define the torsion of $(C^*(M;\rho), \delta^*)$. This torsion is the *Reidemeister torsion* T_{Reid} of (M, ρ).

In fact, T_{Reid} is independent of the triangulation, and so indeed only depends on M and ρ [45]. The Reidemeister torsion can be thought of as a secondary topological invariant; i.e. it is only defined once the primary topological information, namely the cohomology groups of the complex (C^*, δ^*), are trivial. Note that, as promised, the Reidemeister torsion is defined at the cochain level. For the case of lens spaces treated by Franz, the fundamental group is just $\mathbf{Z}_p$ and ρ is one dimensional. Thus ρ takes values in the pth roots of unity, and T_{Reid} is explicitly calculable. As for the teminology, we remark that if ρ is the trivial one dimensional representation, then the Reidemeister torsion (suitably defined for nonexact complexes) is $\exp\sum_q (-1)^q \log |\text{Tor}(H^q(M;\mathbf{Z}))|$.

5.3.2   Analytic Torsion

Given the definition of the Reidemeister torsion, it is not hard (in hindsight) to guess an analytic analogue. Let $E = E_\rho = \tilde{M} \times_\rho \mathbf{C}^N$ be the vector bundle over M associated to ρ: the total space of E_ρ consists of pairs $(\tilde{x}, v) \in \tilde{M} \times \mathbf{C}^N$ modulo the identification $(\tilde{x}, v) \sim (\tilde{y}, w)$ if $\gamma \cdot \tilde{x} = \tilde{y}$ and $\rho(\gamma^{-1})v = w$ for $\gamma \in \pi_1(M)$.

E_ρ comes with a flat connection ∇ (i.e. the curvature of ∇ vanishes). To determine ∇ on a neighborhood U of M over which $\tilde{M}$ equals disjoint copies of U, pick a basis $\{v_i\}$ of $\mathbf{C}^N$ and set $s_i(x)$ to be the equivalence class of $(\tilde{x}, v_i)$, where $\tilde{x}$ is a continuous lift of $x \in U$. Set $\nabla(a^i s_i) = da^i \cdot s_i$. The reader can check that this connection is independent of the various choices and glues together to a connection ∇ on E_ρ, which of course is flat.

We can couple ∇ to the exterior derivative on M to give exterior derivatives d_∇ on the spaces of twisted k-forms $\Lambda^k(M; E_\rho)$. The acyclicity of ρ implies that the de Rham cohomology of the twisted k-forms is trivial. These spaces have inner products coming from a choice of Riemannian metric g on M and the standard metric on $\mathbf{C}^N$, so one can form Laplacians $\Delta^k = d_\nabla^* d_\nabla + d_\nabla d_\nabla^*$. We now wish to define the *analytic torsion* or R-torsion to be

$$T_\rho(M) = T_{\rho,g}(M) = \exp(-\frac{1}{2}\sum_q (-1)^q q \, \log \det \Delta^q).$$

We need to make sense of the term $\det \Delta^q$. As motivation, let $T : V \to V$ be a nonnegative symmetric linear transformation of a finite dimensional inner product space. There exists a basis of eigenvectors for T with eigenvalues $\lambda_1, ..., \lambda_n$. Of course, $\det T = 0$ if any $\lambda_i = 0$. If not, we set $\zeta(s) = \zeta_T(s) = \sum_i (\lambda_i)^{-s}$. It is easy to check that

$$\zeta'(0) = -\log \det T.$$

Now let D be a nonnegative self-adjoint elliptic differential operator acting on sections of a Hermitian bundle E over a compact Riemannian manifold. We have mentioned that such operators have a complete orthonormal set of eigensections on $L^2(E)$, the space of L^2 sections of E, with eigenvalues $\{\lambda_i\}$. As in (5.1), we define the zeta function of D by

$$\zeta(s) = \zeta_D(s) = \sum_{\lambda_i \neq 0} (\lambda_i)^{-s} = \frac{1}{\Gamma(s)} \int_0^\infty t^{s-1} \text{Tr}(e^{-tD} - P) \, dt,$$

where P is the orthogonal projection onto Ker D. We set

$$\det D = \begin{cases} e^{-\zeta'(0)}, & \lambda_i \neq 0, \forall i, \\ 0, & \lambda_i = 0, \text{ some } i. \end{cases}$$

This definition is meaningful, because just as in §5.1 zero is a regular value for the zeta function. This zeta function regularization of $\det D$ is due to Ray and Singer.

Remark: As in Exercise 3, under a scaling of the metric $g \mapsto \lambda^2 g$, $\det D \mapsto \lambda^{\zeta(0)} \det D$. Since the corresponding change in finite dimensions is given by det $T \mapsto \lambda^{\dim V} \det T$, we can view $\zeta(0)$ as a regularized dimension of the space of the (infinite dimensional) domain of D. For example, for the conformal Laplacian, Theorem 5.5 shows that this real valued dimension is independent of the metric in a conformal class. Physicists would say that there is no conformal dimensional anomaly for the conformal Laplacian.

The reader is invited to try to compute $\zeta'(0)$ for the Laplacian on functions using the Mellin transform as in (5.3). One must again break the integral up into $\int_0^1 + \int_1^\infty$, plug the asymptotics for the heat kernel into the first integral, and take $(d/ds)|_{s=0}$. You will find that the first integral contains local terms and an incalculable error term, and the second integral is an intractable non-local expression. Thus a direct approach to calculating $\zeta'(0)$ is not feasible, except in certain cases when the zeta function coincides with known number theoretic zeta functions (e.g. M a symmetric space), where functional equation and representation theoretic techniques are available. The nonlocal nature of $\zeta'(0)$ also demonstrates that analytic torsion is more subtle than the integrals of local expressions that occur in index theory.

Nevertheless, Ray and Singer were able to show that analytic torsion has many key properties of Reidemeister torsion. In particular, analytic torsion depends only on the smooth structure of the manifold (and not on the metric used in its definition), and is trivial on even dimensional manifolds.

Exercise 5: *Show that $T_\rho(M) = 0$ if the dimension n of M is even. Hint: Show that in fact for all s, $\sum_q (-1)^q q \zeta_q(s) = 0$. For as with ordinary forms, $*\Delta^q = \Delta^{n-q}*$ implies that Δ^q, Δ^{n-q} are isospectral. Thus $\zeta_q(s) = \zeta_{n-q}(s)$, which implies that*

$$\sum_q (-1)^q q \zeta_q(s) = \frac{n}{2} \sum_q (-1)^q \zeta_q(s)$$

in even dimensions. Now Lemma 4.1 extended to twisted forms implies that the right hand side of the last equation is zero.

Following [58], we will now prove that the analytic torsion is independent of the metric on an odd dimensional manifold. Fix the representation ρ and a metric g on M, and denote $T_{\rho,g}(M)$ by $T_g(M)$ or just $T(M)$. Since the space of metrics on M is connected by Exercise 7, Chapter 1, it suffices to show that $T_g(M)$ is a smooth function of g with vanishing differential. In other words, given another metric g_1 and a line of metrics $g_\sigma = \sigma g + (1 - \sigma)g_1$, we want to show that

$$\left. \frac{d}{d\sigma} \right|_{\sigma=0} T_{g_\sigma}(M) = 0. \tag{5.13}$$

Since the analytic torsion is defined in terms of the heat kernel trace in the Mellin transform, we first need a result on the variation of the heat kernel trace as a function of σ. Formally, for $\Delta_\sigma = \Delta_\sigma^k$ the Laplacian on twisted k-forms with respect to the metric g_σ, we have

$$\frac{d}{d\sigma}\bigg|_{\sigma=0} \mathrm{Tr}(e^{-t\Delta_\sigma}) = -t\mathrm{Tr}(\dot{\Delta}e^{-t\Delta}),$$

where the dot indicates differentiation with respect to σ and $\Delta = \Delta_0$.

From now on, we will denote $d_\nabla, d_\nabla^* = \delta_\nabla$ just by d, δ. The reader can check that as usual we have $\delta = \pm * d *$. In the definition of Δ, only the star operator depends on the metric. Thus

$$\dot{\Delta} = (d\delta + \delta d)^{\cdot} = \pm d \dot{*} d * \pm d * d \dot{*} \pm \dot{*} d * d \pm * d \dot{*} d.$$

Let α be the endomorphism of k-forms defined by $\alpha = *^{-1}\dot{*}$. The last term in the previous equation equals $\pm * d * *^{-1}\dot{*} d = \delta\alpha d$. The second term similarly equals $d\delta\alpha$. For the first and third terms on the right hand side of the previous equation, we can use the relation

$$*^{-1}\dot{*} = -\dot{*}*^{-1},$$

which follows from $0 = (d/d\sigma)(*^{-1}*)$. We obtain

$$\dot{\Delta} = -d\alpha\delta + d\delta\alpha - \alpha\delta d + \delta\alpha d.$$

In summary, a formal expression for the variation of the heat kernel is

$$\frac{d}{d\sigma}\bigg|_{\sigma=0} \mathrm{Tr}(e^{-t\Delta_\sigma}) = -t\mathrm{Tr}((-d\alpha\delta + d\delta\alpha - \alpha\delta d + \delta\alpha d)e^{-t\Delta}).$$

The following proposition justifies these formal computations.

Proposition 5.14 *The heat kernel $e_\sigma(t, x, y)$ for the metric g_σ is a differentiable function of σ, for each fixed $x, y \in M, t > 0$. We have*

$$\frac{d}{d\sigma}\mathrm{Tr}(e^{-t\Delta_\sigma}) = -t\mathrm{Tr}((\delta\alpha d - d\alpha\delta - \alpha\delta d + d\delta\alpha)e^{-t\Delta_\sigma}).$$

PROOF. (see [58, Prop. 6.1]) For $\sigma, \sigma' \in [0, 1]$, we will denote $\Delta_\sigma, e_\sigma(t, x, y), *_\sigma$, by $\Delta, e(t, x, y), *$, respectively, and similarly we use $\Delta', e'(t, x, y), *'$ for these quantities computed at σ'.

By Exercise 9, Chapter 3, the pointwise trace of the heat kernel is given by $\mathrm{tr}(*_y e(t, x, y)|_{y=x})$, where tr indicates the trace in the fiber $E_x \otimes E_x^*$ with respect to the inner product on E, and $*_y$ indicates the star operator applied in the y variable. Thus we want to compute $\mathrm{tr}(*_y e(t, x, y) - *_y' e'(t, x, y))$, or equivalently $\mathrm{tr}((*_y')^{-1} *_y e(t, x, y) - e'(t, x, y))$, as $\sigma' \to \sigma$. We will need the identities

$$\int_M f \wedge *g = \int_M g \wedge *f = \int_M g \wedge *'(*')^{-1} * f = \int_M (*')^{-1} * f \wedge *'g,$$

which follow from the symmetry of the Hodge inner product, and "Green's formula"

$$\int_M df \wedge *g = \int_M f \wedge *\delta g,$$

which just says that δ is the adjoint of d. Here f, g are E valued forms, and from this point on we are omitting the pointwise trace tr from the notation.

For $*, \Delta$ acting in the u variable, we have

$$(*'_y)^{-1} * _y e(t, x, y) - e'(t, x, y) = \lim_{s \to t} \int_M (*')^{-1} * e(s, x, u) \wedge *'e'(t - s, u, y)$$

$$- \lim_{s \to 0} \int_M e(s, x, u) \wedge *e'(t - s, u, y)$$

$$= \int_0^t ds \frac{\partial}{\partial s} \int_M e(s, x, u) \wedge *e'(t - s, u, y),$$

since the heat kernels approach delta functions as $t \to 0$. By the heat equation, the last expression equals

$$-\int_0^t ds \int_M (\Delta e(s, x, u) \wedge *e'(t - s, u, y) - e(s, x, u) \wedge *\Delta' e'(t - s, u, y)).$$

Now write $\Delta = d\delta + \delta d$ and similarly for Δ'. Applying Green's formula (and the first identity above to handle the $\delta'd$ term), we can transform the previous expression into

$$\int_0^t ds \int_M \Bigg(-de(s, x, u) \wedge *de'(t - s, u, y) - \delta e(s, x, u) \wedge *\delta e'(t - s, u, y)$$

$$+ d((*')^{-1} * e(s, x, u)) \wedge *'de'(t - s, u, y)$$

$$+ \delta e(s, x, u) \wedge *\delta' e'(t - s, u, y) \Bigg).$$

In summary,

$$(*'_y)^{-1} *_y e(t, x, y) - e'(t, x, y)$$

$$= \int_0^t ds \int_M \Bigg(-de(s, x, u) \wedge (* - *')de'(t - s, u, y)$$

$$+ d((*')^{-1}(* - *')e(s, x, u)) \wedge *'de'(t - s, u, y)$$

$$- \delta e(s, x, u) \wedge *(\delta - \delta')e'(t - s, u, y) \Bigg). \qquad (5.15)$$

The estimates of Lemma 4.23 extend to the derivatives $de, \delta e$ etc. of the heat kernel, which also decay exponentially in the distance by the construction of the heat kernel. These estimates show that each term on the right hand side is well behaved near 0 and t, so there is no problem performing the integrations.

In local coordinates, the endomorphism $* - *'$ is given by a matrix with entries depending smoothly on σ, σ'. In particular, there exists a bounded matrix

$A = A(\sigma, \sigma')$ such that $* - *' = (\sigma - \sigma')A(\sigma, \sigma')$, and $\lim_{\sigma' \to \sigma} A(\sigma, \sigma') = \dot{*}(\sigma)$ uniformly on M. This implies that the first term of the right hand side of (5.15) is $O(\sigma - \sigma')$.

The derivatives of the matrix A in fixed local coordinates on M are also bounded and continuous in σ'. Estimates similar to those in Lemma 4.23 can be applied to the terms in (5.15) involving

$$
\begin{aligned}
d((*')^{-1}(* - *')e) &= (\sigma - \sigma')d((*')^{-1}Ae), \\
(\delta - \delta')e &= (\sigma - \sigma')(A *^{-1} \delta + \delta'(*')^{-1}A)e.
\end{aligned}
$$

Thus their contribution to the right hand side of (5.15) is also $O(\sigma - \sigma')$.

Applying the bounded operator $*'_y$ to both sides of (5.15), we see that $*e - *'e'$ is $O(\sigma - \sigma')$ uniformly on M. In particular, $e' = e_{\sigma'}$ is continuous in σ', uniformly on M. By the uniform continuity of A and its derivatives, we get

$$
\begin{aligned}
\frac{d}{d\sigma} *_y (\sigma)e_\sigma(t, x, y) &= \lim_{\sigma' \to \sigma} (\sigma - \sigma')^{-1}(*_y e(t, x, y) - *'_y e'(t, x, y)) \\
&= \int_0^t ds \int_M *_y \{ -de(s, x, u) \wedge \dot{*} de(t - s, u, y) \\
&\quad + d(*^{-1}\dot{*}e(s, x, u)) \wedge *de(t - s, u, y) \\
&\quad - \delta e(s, x, u) \wedge *\dot{*} *^{-1} \delta e(t - s, u, y) \\
&\quad - \delta e(s, x, u) \wedge *\delta(*^{-1}\dot{*}e(t - s, u, y)) \}.
\end{aligned}
$$

The integrations are justified as above. We see that $(d/d\sigma) * e$ satisfies the same exponential decay estimate as e. This allows us to apply Green's formula (after replacing $\dot{*}de(t - s, u, y)$ in the first term on the right hand side with $* *^{-1} \dot{*}de(t - s, u, y)$) to obtain

$$
\begin{aligned}
\frac{d}{d\sigma} *_y (\sigma)e_\sigma(t, x, y) &= \int_0^t ds \int_M *_y \{ -e(s, x, u) \wedge *\delta *^{-1} \dot{*}de(t - s, u, y) \\
&\quad + *^{-1}\dot{*}e(s, x, u) \wedge *\delta de(t - s, u, y) \\
&\quad - e(s, x, u) \wedge *d\dot{*} *^{-1} \delta e(t - s, u, y) \\
&\quad + *e(s, x, u) \wedge *d\delta *^{-1} \dot{*}e(t - s, u, y) \}.
\end{aligned}
$$

The second term in the last expression can be transformed using

$$
\begin{aligned}
\int_M *^{-1}\dot{*}f \wedge *g &= \int_M g \wedge \dot{*}f = \frac{d}{d\sigma} \int_M g \wedge *f = \frac{d}{d\sigma} \int_M f \wedge *g \\
&= \int_M f \wedge \dot{*}g = \int_M f \wedge * *^{-1} \dot{*}g
\end{aligned}
$$

(which says that α is self-adjoint) and $\dot{*}*^{-1} = -*^{-1}\dot{*}$. The result is the remarkable formula

$$
\frac{d}{d\sigma} *_y (\sigma)e_\sigma(t, x, y) = -\int_0^t ds \int_M *_y \{ e(s, x, u) \wedge *\delta \alpha de(t - s, u, y)
$$

$$-e(s,x,u) \wedge *\alpha\delta de(t-s,u,y)$$
$$-e(s,x,u) \wedge *d\alpha\delta e(t-s,u,y)$$
$$+e(s,x,u) \wedge *d\delta\alpha e(t-s,u,y)\}$$
$$= -\int_0^t ds \int_M *_u\{(\delta\alpha d - \alpha\delta d - d\alpha\delta + d\delta\alpha)_u$$
$$\times e(s,x,v) \wedge *_y e(t-s,u,y)\}|_{v=u}. \qquad (5.16)$$

To obtain

$$\frac{d}{d\sigma}\mathrm{Tr}(e^{-t\Delta_\sigma}) = \int_M \frac{d}{d\sigma}*_y e_\sigma(t,x,y)|_{y=x},$$

we must set $y = x$ and integrate the right hand side of (5.16) over M with respect to x. In doing this, we can interchange the order of integration, and integrate with respect to x before applying d, δ, α to the u variable. Each of the integrals with respect to x reduces to

$$\int_M e(s,x,v) \wedge *_x e(t-s,u,x) = e(t,u,v),$$

by the semigroup property of the heat kernel. Hence $(d/d\sigma) \, \mathrm{Tr} \, (e^{-t\Delta_\sigma})$ is given by the integral over M of

$$-t *_u (\delta\alpha d - \alpha\delta d - d\alpha\delta + \alpha d\delta)_u e(t,u,v)$$

taken at $v = u$. This finishes the proof.

We can now show that analytic torsion is independent of the metric on an odd dimensional manifold. We use the notation of (5.13).

Theorem 5.17 (Ray-Singer) *If the dimension of M is odd, then*

$$\frac{d}{d\sigma}\bigg|_{\sigma=0} T_{g_\sigma}(M) = 0.$$

In particular, analytic torsion is independent of the metric on M, and so defines a smooth invariant of the pair (M, ρ).

PROOF. Since ρ is acyclic, Ker $\Delta_\sigma^k = 0$ for all k, σ. (The proof of this statement involves a straightforward extension of the Hodge theorem for forms to forms with values in E.) Thus the zeta function on E valued k-forms is given by

$$\zeta_k(s) = \frac{1}{\Gamma(s)} \int_0^\infty t^{s-1} \mathrm{Tr}(e^{-t\Delta^k}) \, dt.$$

As in (5.7), it suffices to show that

$$f(\sigma, s) = \frac{1}{2}\sum_{q=0}^N (-1)^q q \int_0^\infty t^{s-1} \mathrm{Tr}(e^{-t\Delta_\sigma^q}) \, dt$$

has $(\partial/\partial\sigma)f(\sigma,0) = 0$. Of course, this means that $f(\sigma,s)$, which exists for $\mathrm{Re}(s)$ sufficiently large, has a meromorphic continuation to $\mathbf{C}$ with the required derivative vanishing at 0.

The Rayleigh-Ritz characterization of the first eigenvalue of a positive operator Δ on a Hilbert space $\mathcal{H}$,

$$\lambda_1 = \inf_{f \in \mathcal{H}} \frac{\langle \Delta f, f \rangle}{\langle f, f \rangle},$$

shows that λ_1 for Δ_σ is a continuous function of σ. Thus for any fixed $t_0 > 0$, there exist C, $\epsilon > 0$ independent of $\sigma \in [0,1]$ such that $\mathrm{Tr}(e^{-t\Delta_\sigma^q}) \leq C e^{-\epsilon t}$ for all $t > t_0$. The previous proposition then implies that there exist different C, ϵ such that

$$\frac{\partial}{\partial\sigma}\mathrm{Tr}(e^{-t\Delta_\sigma^q}) \leq C e^{-\epsilon t}$$

independently of σ. Thus we may differentiate under the integral to get

$$\frac{\partial}{\partial\sigma}f(\sigma,s) = -\frac{1}{2}\sum_{q=0}^{N}(-1)^q q \int_0^\infty t^s \mathrm{Tr}((\dot{\Delta}_\sigma^q)\, e^{-t\Delta_\sigma^q})dt \qquad (5.18)$$

for $\mathrm{Re}(s)$ large.

We now compute $\mathrm{Tr}(\dot{\Delta}^q e^{t\Delta_q})$. If A is of trace class (i.e. the sum of its eigenvalues is finite) and B is a bounded operator, it is well known that AB, BA are of trace class and $\mathrm{Tr}(AB) = \mathrm{Tr}(BA)$. Now the operators $\alpha\delta d$, etc. in $\dot{\Delta}^q$ are not bounded, but $\alpha\delta d e^{-t\Delta^q}$, etc. are bounded operators on L^2. We can use the semigroup property of the heat operator to write

$$\begin{aligned}
\mathrm{Tr}(\delta\alpha d e^{-t\Delta^q}) &= \mathrm{Tr}(\delta\alpha d e^{-\frac{1}{2}\Delta^q}e^{-\frac{1}{2}\Delta^q}) = \mathrm{Tr}(e^{-\frac{1}{2}\Delta^q}\delta\alpha d e^{-\frac{1}{2}\Delta^q}) \\
&= \mathrm{Tr}(\alpha d e^{-\frac{1}{2}\Delta^q}e^{-\frac{1}{2}\Delta^q}\delta) = \mathrm{Tr}(\alpha d e^{-t\Delta^q}\delta) \\
&= \mathrm{Tr}(\alpha d\delta e^{-t\Delta^{q+1}}),
\end{aligned}$$

where we have used $\Delta^q\delta = \delta\Delta^{q+1}$ in the last step. Similarly, we get

$$\mathrm{Tr}(d\alpha\delta e^{-t\Delta^q}) = \mathrm{Tr}(\alpha d\delta e^{-t\Delta^{q-1}}), \ \mathrm{Tr}(d\delta\alpha e^{-t\Delta^q}) = \mathrm{Tr}(\alpha d\delta e^{-t\Delta^q}),$$

using $\Delta^q d = d\Delta^{q-1}$. As a result,

$$\begin{aligned}
\mathrm{Tr}(\dot{\Delta}^q e^{t\Delta^q}) &= -\mathrm{Tr}(\alpha\delta d e^{t\Delta^q}) + \mathrm{Tr}(\alpha d\delta e^{t\Delta^{q+1}}) \\
&\quad -\mathrm{Tr}(\alpha\delta d e^{t\Delta^{q-1}}) + \mathrm{Tr}(\alpha d\delta e^{t\Delta^q}).
\end{aligned}$$

Thus we get a telescoping sum

$$\begin{aligned}
\sum_{q=0}^{N}(-1)^q q\,\mathrm{Tr}(\dot{\Delta}^q e^{-t\Delta^q}) &= \sum_{q=0}^{N}(-1)^q(\mathrm{Tr}(\alpha\delta d e^{-t\Delta^q}) + \mathrm{Tr}(\alpha d\delta e^{-t\Delta^q})) \\
&= \sum_{q=0}^{N}(-1)^q\,\mathrm{Tr}(\alpha\Delta e^{-t\Delta^q}) \\
&= -\frac{d}{dt}\sum_{q=0}^{N}(-1)^q\,\mathrm{Tr}(\alpha e^{-t\Delta^q}).
\end{aligned}$$

Plugging this back into (5.18) gives

$$\frac{\partial}{\partial \sigma} f(\sigma, s) = \frac{1}{2} \sum_{q=0}^{N} (-1)^q \int_0^\infty t^s \frac{d}{dt} \mathrm{Tr}(\alpha e^{-t\Delta_\sigma^q}) \, dt$$

$$= -\frac{1}{2} s \sum_{q=0}^{N} (-1)^q \int_0^\infty t^{s-1} \mathrm{Tr}(\alpha e^{-t\Delta_\sigma^q}) \, dt \qquad (5.19)$$

by integration by parts. The boundary terms in the integration by parts vanish as in the proof of Theorem 5.5. As in Exercise 4,

$$\mathrm{Tr}(\alpha e^{-t\Delta_\sigma^q}) \sim \left(\sum_{k=0}^{\infty} \int_M \mathrm{tr}(\alpha u_k^q) \right) t^{k-(n/2)}.$$

As in Theorem 5.2, this shows that

$$s \int_0^\infty t^{s-1} \mathrm{Tr}(\alpha e^{-t\Delta_\sigma^q}) \, dt$$

has a meromorphic continuation to $\mathbf{C}$ whose value at zero is

$$\begin{cases} \int_M \mathrm{tr}(\alpha u_{n/2}), & \dim M \text{ even}, \\ 0, & \dim M \text{ odd} \end{cases}$$

(since s and $1/\Gamma(s)$ both have a simple zero at $s = 0$). Note that this uses the acyclicity of ρ. By (5.19), the meromophic continuation of $(\partial/\partial\sigma)f(\sigma, s)$ vanishes at $s = 0$ if $\dim M$ is odd.

The fact that analytic torsion is a smooth invariant of a manifold, as well at its other properties, led Ray and Singer to conjecture the equality of Reidemeister and analytic torsion. The various proofs known of this equality are too difficult to include here, so we just mention the basic techniques used. In Cheeger's proof, the behavior of analytic torsion under surgery is compared to the known behavior of Reidemeister torsion under surgery. After a sequence of surgeries, the equality of the two torsions is reduced to their known equality on the sphere. In Müller's proof, the convergence of the eigenvalues of the combinatorial Laplacians to the eigenvalues of the smooth Laplacians as the triangulation gets finer and finer is shown, which, combined with surgery techniques, provides the necessary link between the combinatorial Reidemeister torsion and the analytic torsion of the smooth structure. In a recent proof of Vishik [66], the surgery argument is combined with a "gluing formula" for the two torsions motivated by topological field theory. Finally, in the Bismut-Zhang proof, both torsions are related to the torsion of the Morse-Smale complex associated to a Morse function on the manifold. The localization of the infinite dimensional analytic information onto the finite dimensional Morse-Smale complex generalizes Witten's analytic proof of the Morse inequalities ([69], [18], [59]). This proof has the advantage of taking place on one fixed manifold, although with the additional data of a Morse function; it is also the most technically difficult proof.

5.3.3 The Families Index Theorem and Analytic Torsion

So far in this section, we have had several indications that torsion is a "secondary" invariant, one that can only (easily) be defined once the relevant "primary" cohomological invariants vanish. This vanishing is clearly necessary in our definition of Reidemeister torsion, and is necessary in order for the analytic torsion to be independent of the Riemannian metric. Moreover, as we showed in §5.1, $\zeta'(0)$ for the conformal Laplacian is a nontrivial conformal invariant only when $\zeta(0)$ vanishes, and $\zeta(0)$ is a combination of a locally computable, index theoretic term and a cohomology-like term. Thus we expect the two torsions to have significance precisely where index theory fails to give information; indeed, torsion is nontrivial only on odd dimensional manifolds, where index theory is trivial.

In summary, there should be some relationship between torsion and index theory, some mathematical object which is the torsion when the appropriate cohomology vanishes. Such a relationship does exist by recent work of Bismut and Lott [8].

To put their work in a historical context, we first discuss the index theorem for families of elliptic operators. Roughly speaking, this theorem measures the change of the kernel and cokernel of a family of elliptic operators, parametrized by a smooth manifold B, acting on fixed bundles over a fixed manifold Z. As in Chapter 4, we will restrict attention to families of twisted signature operators.

We consider a fibration $Z \to M \xrightarrow{\pi} B$, with Z and B (and hence M) smooth manifolds, together with a bundle E over M and a family of twisted signature operators $\{D_b : b \in B\}$ acting on $\Lambda^* T^*(Z_b) \otimes E|_{Z_b}$, where $Z_b = \pi^{-1}(b)$ is the fiber over b. Of course, to define these operators we need to have smoothly varying Riemannian metrics on each Z_b, and we assume that Z, M, B are compatibly oriented. Associated to each fiber we have two finite dimensional subspaces of $\Lambda^* T^* Z_b \otimes E|_{Z_b}$, namely Ker D_b and Ker D_b^*. Assuming these spaces have dimension independent of $b \in B$, we obtain two bundles, denoted KER D and KER D^*, over B. (Even if the dimensions of the kernel and cokernel jump, the virtual bundle IND D whose fiber is the formal difference Ker D_b − Ker D_b^* is well defined, see [2]. IND D is called the *index bundle* of the family D.) It is known that complex vector bundles F are topologically classified up to torsion by their Chern character ch(F); more precisely, the map ch : $K(B) \otimes \mathbf{Q} \to H^{\mathrm{ev}}(B; \mathbf{Q})$ is an isomorphism from the torsion free part of the ring $K(B)$ of virtual bundles over B to the even dimensional rational cohomology of B. Granted this, we see that the twisting of the kernel and cokernel of D_b is captured by ch(KER D) − ch(KER D^*), or more simply by ch(IND D), if we extend ch to a ring homomorphism of $K(B)$. To bring things down to earth, the reader should check that if $B = \{b\}$ is a point, then ch(KER D) − ch(KER D^*) is just the index of the single operator D_b.

The Atiyah-Singer families index theorem [2] identifies the element ch(IND D) in $H^{\mathrm{ev}}(B; \mathbf{Q})$. In keeping with the heat equation approach of the last chapter, we will state a more refined result in terms of differential forms; the first heat equation proof was given by Bismut in [6].

To set the notation, define the tangent bundle along the fibers to be $TZ \overset{\text{def}}{=}$ Ker π_*. This is a bundle over M, which comes with a metric g^{TZ} built from the family of metrics on Z. Choose a horizontal distribution $T^H M \subset TM$; i.e. a smooth distribution in TM with $TZ \oplus T^H M = TM$. A choice of metric g^B on B gives rise to a metric $g^M = g^{TZ} \oplus \pi^* g^B$ on M, under the convention that $T^H M$ is orthogonal to TZ. The Levi-Civita connection ∇^M for g^M induces a connection ∇^{TZ} on TZ by $\nabla^{TZ} = P^{TZ}\nabla^M$, where $P^{TZ} : TM \to TZ$ is orthogonal projection. (In fact, ∇^{TZ} is independent of the choice of metric g^B.) Denote the curvature of ∇^{TZ} by Ω^{TZ}. Choose a connection ∇^E on E with curvature Ω^E. Assume the dimension of M is $2k$, and let L_k be the kth Hirzebruch polynomial. Finally, we must define $\int_M : \Lambda^* T^* M \to \Lambda^{*-2k} T^* B$, the operation of integration along the fibers, which takes a q-form ω on M to a $(q - 2k)$-form $\int_M \omega$ on B. Writing ω locally as $\omega(b, z) = a_{IJ}(b, z)db^I \wedge dz^J$, where b, z are local coordinates for B, Z respectively, we set

$$\left(\int_M \omega \right)(z) = \sum_{J, |J|=2k} \left(\int_{Z_b} a_{IJ}(b, z)dz^J \right) db^I,$$

where the right hand side integral is ordinary integration of forms.

Note that the space of sections $\Gamma(\Lambda^* T^* Z_b \otimes E|_{Z_b})$ has an L^2 inner product induced by the metric on the fibers and the metric on E. Considering these spaces as forming an infinite dimensional Hermitian bundle $\mathbf{E}$ over B, Bismut constructs a unitary (super)connection $\nabla^{\mathbf{E}}$ on $\mathbf{E}$. The finite dimensional subbundles KER D, KER D^* inherit connections $\nabla^\pm = P^\pm \nabla^{\mathbf{E}}$, where $P^\pm$ denotes orthogonal projection from $\Gamma(\Lambda^* T^* Z_b \otimes E|_{Z_b})$ to Ker D_b, Ker D_b^*, respectively. These connections have curvature $\Omega^\pm$, so the Chern character ch(IND D) of the index bundle is represented in de Rham cohomology by the even dimensional form ch(Ω^+) − ch(Ω^-).

With all this notation, we can finally state the index theorem for families.

Theorem 5.20 (The Families Index Theorem) *The de Rham cohomology class of* ch(IND D) *in* $H^{\text{ev}}(B; \mathbf{Q})$ *is given by*

$$[\text{ch}(\Omega^+) - \text{ch}(\Omega^-)] = \left[\int_M L_k(\Omega^{TZ})\text{ch}(\Omega^E) \right]. \tag{5.21}$$

Notice that the right hand side is a sum of even dimensional forms on B which are locally computable from curvature information on M and E. The reader can check that these forms are closed, and so give real (in fact rational) cohomology classes on B. In fact, Bismut constructs an odd dimensional form α such that

$$\text{ch}(\Omega^+) - \text{ch}(\Omega^-) = \int_M L_k(\Omega^{TZ})\text{ch}(\Omega^E) + d\alpha. \tag{5.22}$$

Thus (5.21) can be refined to an equality of differential forms, and so gives a local geometric version of the original cohomological formulation of the families index theorem due to Atiyah-Singer. Bismut's proof follows the arguments of

Chapter 4; he constructs a heat-type operator whose (super)trace approaches the left hand side of (5.22) as $t \to 0$ and the right hand side as $t \to \infty$. A survey of recent applications of the families index theorem is in [60].

We now assume that E is flat; i.e. E has a metric $h = h^E$ and an h-unitary connection ∇^E whose curvature satisfies $\Omega^E = 0$. Locally there exists a basis $\{e_i\}$ of the fibers of E with $\nabla^E e_i = 0$. With respect to this basis, h is a function on M with values in Hermitian matrices. Define an $\mathrm{End}(E)$ valued one-form on M by $\omega(E, h) = h^{-1}dh$. For odd positive integers k, set

$$c_k(E, h) = (2\pi i)^{-(k-1)/2} 2^{-k} \mathrm{Tr}(\omega^k(E, h)),$$

where ω^k is ω wedged with itself k times. The k-form $c_k(E, h)$ is closed and its cohomology class $c_k(E) \in H^k(M; \mathbf{R})$ is independent of h. Let $e(TZ)$ be the Euler class of the bundle TZ; for our purposes, this can be constructed by taking a unitary connection ∇^{TZ} on TZ, and then forming the cohomology class $e(TZ)$ of the Pfaffian $\mathrm{Pf}(\Omega^{TZ})$ of the curvature Ω^{TZ}. Let $H^p(Z; E|_Z)$ denote the flat vector bundle over B whose fiber at $b \in B$ is the cohomology group $H^p(Z_b; E|_{Z_b})$. Finally, we remark that integration along the fiber takes closed forms on M to closed forms on B, and exact forms to exact forms, and hence induces a map $\int_Z : H^*(M; \mathbf{R}) \to H^{*-\dim Z}(B; \mathbf{R})$ (see [11]). Then we have the following cohomological result due to Bismut and Lott.

Theorem 5.23 *For any odd integer k,*

$$\int_Z e(TZ)c_k(E) = \sum_{p=0}^{\dim Z} (-1)^p c_k(H^p(Z; E|_Z)) \qquad (5.24)$$

as classes in $H^k(B; \mathbf{R})$.

We remark that Theorem 5.23 is a C^∞ analogue of a local version of the Grothendieck-Riemann-Roch theorem for holomorphic submersions due to Bismut, Gillet and Soulé [7], and as such is a families-type index theorem. Our interest is more in the restatement of Theorem 5.23 at the level of differential forms. As in the families index theorem, the fibers $H^p(Z_b; E|_{Z_b})$ inherit an inner product h^{H^p} from the L^2 inner product on $\Gamma(TZ_b \otimes E|_{Z_b})$ (since by Hodge theory $H^p(Z_b; E|_{Z_b})$ is isomorphic to a space of harmonic forms in $L^2(TZ_b \otimes E|_{Z_b})$). Recall that the metric on M depended on a choice of the horizontal distribution $T^H M$.

Theorem 5.25 (Bismut-Lott) *There exist $(k-1)$-forms*

$$\mathcal{T}_{k-1} = \mathcal{T}_{k-1}(T^H M, g^{TZ}, h^E)$$

on B such that
(i) for any odd integer k,

$$d\mathcal{T}_{k-1} = \int_Z \mathrm{Pf}(\Omega^{TZ})c_k(E, h^E) - \sum_{p=0}^{\dim Z} (-1)^p c_k(H^p(Z; E|_Z), h^{H^p}) \qquad (5.26)$$

as differential k-forms;

(ii) if the cohomology groups $H^p(Z_b; E|_{Z_b})$ vanish for all p and all $b \in B$, and if Z is odd dimensional, then the forms $\mathcal{T}_{k-1}$ are closed, and the class $[\mathcal{T}_{k-1}] \in H^{k-1}(B; \mathbf{R})$ is independent of the choices of $T^H M, g^{TZ}, h^E$. In particular, $[\mathcal{T}_0]$ is (represented by) the locally constant function which is half the analytic/Reidemeister torsion of the pair $(Z_b, E|_{Z_b})$.

This result has several striking features. First, it realizes the analytic/Reidemeister torsion as one element of a sophisticated local familes-type index theorem. Second, note that the first statement in (ii) follows immediately from (i), since dim Z odd implies $\int_Z \mathrm{Pf}(\Omega^{TZ}) c_k(E, h^E)$ is an even form and so has no component in degree k. Thus, as expected, the *higher torsion forms* $\mathcal{T}_{k-1}$ have cohomological significance only when the cohomological information in the index-type Theorem 5.23 is trivial. Third, if dim Z is odd and $k = 1$, then (5.26) precisely measures the dependence of the analytic torsion of $(Z_b, E|_{Z_b})$ on the metrics parametrized by B. This dependence is nontrivial if the cohomology groups $H^p(Z_b; E|_{Z_b})$ are nontrivial.

Thus the higher torsion forms are the appropriate mathematical objects, mentioned in the beginning of this subsection, which provide geometric information in general and topological information when index theory is trivial. The existence of the *higher torsion classes* $[\mathcal{T}_{k-1}]$ under the hypotheses in (ii) was conjectured by Wagoner [67], who was motivated by considerations in algebraic K-theory. It is striking that these classes were first uncovered by analytic methods. A proposed topological definition of the higher torsion classes is in [34], [39], but it is unknown at present if the topological and analytic definitions coincide. Together with other work relating index theory to algebraic K-theory [12], [36], [42], the Bismut-Lott result points towards a deep, as yet uncharted, relationship between the two fields.

Bibliography

[1] M. F. Atiyah, R. Bott, and V. K. Patodi, *On the heat equation and the index theorem*, Inventiones Math. **19** (1973), 279–330.

[2] M. F. Atiyah and I. M. Singer, *The index of elliptic operators. III*, Ann. Math. **87** (1968), 546–604.

[3] T. Aubin, *Nonlinear Analysis on Manifolds. Monge-Ampère Equations*, Grundlehren der mathematischen Wissenschaften, vol. 252, Springer-Verlag, Berlin, 1982.

[4] M. Berger, P. Gauduchon, and E. Mazet, *Le Spectre d'une variété riemannienne*, Lecture Notes in Mathematics, vol. 194, Springer-Verlag, Berlin, 1971.

[5] N. Berline, E. Getzler, and M. Vergne, *Heat Kernels and Dirac Operators*, Grundlehren der mathematischen Wissenschaften, vol. 298, Springer-Verlag, Berlin, 1992.

[6] J.-M. Bismut, *The index theorem for families of Dirac operators: two heat equation proofs*, Inventiones Math. **83** (1986), 91–151.

[7] J.-M. Bismut, H. Gillet, and C. Soulé, *Analytic torsion and holomorphic determinant bundles. I–III*, Commun. Math. Phys. **115** (1988), 49–78, 79–126, 301–351.

[8] J.-M. Bismut and J. Lott, *Flat vector bundles, direct images, and higher real analytic torsion*, J. Amer. Math. Soc. **8** (1995), 291–363.

[9] J.-M. Bismut and W. Zhang, *An extension of a theorem of Cheeger and Müller*, Astérisque **205** (1992), 3–235.

[10] A. Borel, *Compact Clifford-Klein forms for symmetric spaces*, Topology **2** (1963), 111–122.

[11] R. Bott and L. Tu, *Differential Forms in Algebraic Topology*, Graduate Texts in Mathematics, vol. 82, Springer-Verlag, Berlin, 1982.

[12] U. Bunke, *Higher analytic torsion and cohomology of diffeomorphism groups*, preprint, http://www.uni-math.gwdg.de/bunke/project3.html, 1998.

[13] P. Buser, *Geometry and Spectra of Compact Riemann Surfaces*, Progress in Mathematics, vol. 106, Birkhäuser, Boston, Mass., 1992.

[14] J. Cao, *Certain 4-manifolds with non-negative curvature*, preprint, 1993.

[15] J. Cheeger, *Analytic torsion and the heat equation*, Ann. Math. **109** (1979), 259–322.

[16] J. Cheeger and S.-T. Yau, *A lower bound for the heat kernel*, Commun. Pure Appl. Math. **34** (1981), 465–480.

[17] S.-S. Chern, *A simple intrinsic proof of the Gauss-Bonnet theorem for closed Riemannian manifolds*, Ann. Math. **45** (1944), 747–752.

[18] H. Cycon, R. Froese, W. Kirsch, and B. Simon, *Schrödinger Operators with Applications to Quantum Mechanics and Global Geometry*, Springer-Verlag, Berlin, 1987.

[19] G. de Rham, *Differentiable Manifolds: Forms, Currents, Harmonic forms*, Grundlehren der mathematischen Wissenschaften, vol. 266, Springer-Verlag, Berlin, 1984.

[20] J. Dieudonné, *A History of Algebraic and Differential Topology 1900–1960*, Birkhäuser, Boston, Mass., 1989.

[21] M. do Carmo, *Differential Geometry of Curves and Surfaces*, Prentice–Hall, Englewood Cliffs, NJ, 1976.

[22] S. Donaldson, *An application of gauge theory to the topology of 4-manifolds*, J. Differential Geometry **18** (1983), 269–316.

[23] K. D. Elworthy, *Geometric aspects of diffusions on manifolds*, École d'Été de Probabilités de Saint–Flour XV–XVII, 1985–87 (P. L. Hennequin, ed.), Lecture Notes in Mathematics, vol. 1362, Springer-Verlag, 1988, pp. 277–426.

[24] G. B. Folland, *Introduction to Partial Differential Equations*, Princeton University Press, Princeton, NJ, 1995.

[25] D. Freed and K. Uhlenbeck, *Instantons and Four-Manifolds*, Springer-Verlag, New York, 1984.

[26] M. Gaffney, *Hilbert space methods in the theory of harmonic integrals*, Trans. Amer. Math. Soc. **78** (1955), 426–444.

[27] S. Gallot, D. Hulin, and J. Lafontaine, *Riemannian Geometry*, Springer-Verlag, Berlin, 1987.

[28] E. Getzler, *A short proof of the local Atiyah-Singer index theorem*, Topology **25** (1986), 111–117.

[29] P. B. Gilkey, *Curvature and the eigenvalues of the Laplacian for elliptic complexes*, Adv. in Math. **10** (1973), 344–381.

[30] _____, *Invariance Theory, the Heat Equation, and the Atiyah-Singer Index Theorem*, Publish or Perish, Wilmington, Del., 1984.

[31] M. Gromov, *Structures métriques sur les variétés riemanniennes*, Cedic, Paris, 1981.

[32] V. Guillemin and A. Pollack, *Differential Topology*, Prentice-Hall, Englewood Cliffs, NJ, 1974.

[33] F. Hirzebruch, *Automorphe Formen und der Satz von Riemann-Roch*, Symposium Internacional de Topologia Algebraica, UNESCO, 1958, pp. 129–144.

[34] K. Igusa, *Parametrized Morse theory and its applications*, Proc. Int. Cong. Math., Kyoto 1990, Mathematical Society of Japan, Tokyo (1991), 643–651.

[35] A. Ikeda, *On lens spaces which are isospectral but not isometric*, Ann. Sci. École Norm. Sup. **13** (1980), 303–315.

[36] J. D. S. Jones and B. Westbury, *Algebraic K-theory, homology spheres, and the η-invariant*, preprint, 1993.

[37] M. Kac, *Can you hear the shape of a drum?*, Amer. Math. Monthly **73** (1966), 1–23.

[38] J. Kazdan and F. Warner, *Existence and conformal deformation of metrics with prescribed Gaussian and scalar curvature*, Ann. Math. **101** (1975), 317–331.

[39] J. Klein, *Higher Franz-Reidemeister torsion: low-dimensional applications*, Contemporary Mathematics **150** (1993), American Mathematical Society, Providence, RI, pps. 195–204.

[40] S. Kudla and J. Millson, *Harmonic differentials and closed geodesics on a Riemann surface*, Invent. Math. **54** (1979), 193–211.

[41] J. Lohkamp, *Negatively Ricci curved manifolds*, Bull. Amer. Math. Soc. **27** (1992), 288–291.

[42] J. Lott, *Diffeomorphisms, analytic torsion and noncommutative geometry*, dg-ga/9607006.

[43] H. McKean and I. M. Singer, *Curvature and the eigenvalues of the Laplacian*, J. Differential Geometry **1** (1967), 43–69.

[44] A. Milgram and P. Rosenbloom, *Harmonic forms and heat conduction, I, II*, Proc. Nat. Acad. Sci. USA **37** (1951), 180–184, 435–438.

[45] J. Milnor, *Whitehead torsion*, Bull. Amer. Math. Soc. **72** (1966), 358–426.

[46] ———, *Morse Theory*, Annals of Mathematics Study, vol. 51, Princeton University Press, Princeton, NJ, 1973.

[47] J. Milnor and J. D. Stasheff, *Characteristic Classes*, Annals of Mathematics Study, vol. 76, Princeton University Press, Princeton, NJ, 1974.

[48] S. Minakshisundaram and A. Pleijel, *Some properties of the eigenfunctions of the Laplace operator on Riemannian manifolds*, Can. J. Math. **1** (1949), 242–256.

[49] S. Molchanov, *Diffusion processes and Riemannian geometry*, Russ. Math. Surveys **30** (1975), 1–63.

[50] G. Mostow and Y.-T. Siu, *A compact Kähler surface not covered by the ball*, Ann. Math. **112** (1980), 321–360.

[51] W. Müller, *Analytic torsion and R-torsion of Riemannian manifolds*, Adv. in Math. **28** (1978), 233–305.

[52] T. Parker, *Heat convolutions and path space*, preprint, 1985.

[53] ———, *Supersymmetry and the Gauss-Bonnet theorem*, preprint, 1985.

[54] T. Parker and S. Rosenberg, *Invariants of conformal Laplacians*, J. Differential Geometry **25** (1987), 535–557.

[55] V. K. Patodi, *An analytic proof of the Riemann-Roch-Hirzebruch theorem*, J. Differential Geometry **5** (1971), 251–283.

[56] ———, *Curvature and the eigenforms of the Laplacian*, J. Differential Geometry **5** (1971), 233–249.

[57] M. Protter and H. Weinberger, *Maximum Principles in Differential Equations*, Prentice-Hall, Englewood Cliffs, NJ, 1967.

[58] D. B. Ray and I. M. Singer, *R-torsion and the Laplacian on Riemannian manifolds*, Adv. in Math. **7** (1971), 145–210.

[59] J. Roe, *Elliptic Operators, Topology, and Asymptotic Methods*, Pitman Research Notes in Mathematics, vol. 179, Longman Scientific and Technical, Burnt Mill, UK, 1988.

[60] S. Rosenberg, *Nonlocal invariants in index theory*, Bulletin AMS **34** (1997), 423–434.

[61] R. Seeley, *Complex powers of an elliptic operator*, Proc. Symp. Pure Math. **10** (1967), 288–307.

[62] P. Shanahan, *The Atiyah-Singer Index Theorem: An Introduction*, Lecture Notes in Mathematics, vol. 638, Springer-Verlag, Berlin, 1987.

[63] M. Spivak, *Calculus on Manifolds*, Addison-Wesley, Redwood City, Calif., 1965.

[64] _____, *A Comprehensive Introduction to Differential Geometry*, vol. I, II, IV, Publish or Perish, Boston, Mass., 1975.

[65] T. Sunada, *Riemannian coverings and isospectral manifolds*, Ann. of Math. **121** (1985), 169–186.

[66] S. Vishik, *Generalized Ray-Singer conjecture. I. A manifold with a smooth boundary*, Commun. Math. Phys. **167** (1995), 1–102.

[67] J. Wagoner, *Diffeomorphisms, K_2 and analytic torsion*, Proc. Symp. Pure and Appl. Math. **39** (1978), 23–33.

[68] F. Warner, *Foundations of Differentiable Manifolds and Lie Groups*, Scott, Foresman and Company, Glenview, Ill., 1971.

[69] E. Witten, *Supersymmetry and Morse theory*, J. Differential Geometry **121** (1985), 169–186.

[70] K. Yoshida, *Functional Analysis*, 6th ed., Springer-Verlag, Berlin, 1980.

Index

Printed in the United States
By Bookmasters